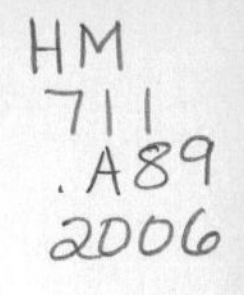

THE EVOLUTION OF COOPERATION

Robert Axelrod

A Member of the Perseus Books Group

OCM 76963800
Santiago Canyon College
Library

An earlier version of chapter 2, tables 1-5, and figures 1-2 appeared in *The Journal of Conflict Resolution* (1980).

An earlier version of chapter 3 appeared in the *American Political Science Review* (1981).

An earlier version of chapter 5 appears in *Science* 211 (27 March 1981):1390-96.
Copyright 1981 by the American Association for the Advancement of Science.

Copyright © 1984 by Robert Axelrod
Originally published in 1984 by Basic Books,
A Member of the Perseus Books Group
Revised edition published in 2006 by Basic Books

All rights reserved. Printed in the United States of America. No part of this book may be reproduced in any manner whatsoever without written permission except in the case of brief quotations embodied in critical articles and reviews. For information, address Basic Books, 387 Park Avenue South, New York, NY 10016–8810.

Books published by Basic Books are available at special discounts for bulk purchases in the United States by corporations, institutions, and other organizations. For more information, please contact the Special Markets Department at the Perseus Books Group, 11 Cambridge Center, Cambridge MA 02142, or call (617) 252-5298 or (800) 255-1514, or e-mail special.markets@perseusbooks.com.

A CIP catalog record for this book is available from the Library of Congress
Original publication: 0-465-02122-0 (cloth); 0-465-02121-2 (paper)
Revised edition:
ISBN-13: 978-0-465-00564-2
ISBN-10: 0-465-00564-0

10 9 8 7 6 5 4 3 2 1

CONTENTS

PART IV
Advice for Participants and Reformers

PART V
Conclusions

APPENDIX A

APPENDIX B

PREFACE

THIS PROJECT began with a simple question: When should a person cooperate, and when should a person be selfish, in an ongoing interaction with another person? Should a friend keep providing favors to another friend who never reciprocates? Should a business provide prompt service to another business that is about to be bankrupt? How intensely should the United States try to punish the Soviet Union for a particular hostile act, and what pattern of behavior can the United States use to best elicit cooperative behavior from the Soviet Union?

There is a simple way to represent the type of situation that gives rise to these problems. This is to use a particular kind of game called the iterated Prisoner's Dilemma. The game allows the players to achieve mutual gains from cooperation, but it also allows for the possibility that one player will exploit the other, or the possibility that neither will cooperate. As in most realistic situations, the players do not have strictly opposing interests. To find a good strategy to use in such situations, I invited experts in game theory to submit programs for a Computer Prisoner's Dilemma Tournament—much like a computer chess tournament. Each program would have available to it the history of the interaction so far and could use this history in making its choice of whether or not to cooperate on the current move. Entries came from game theorists in economics, psychology, sociology, political science, and mathematics. I ran the fourteen entries and a random rule against each other in a round robin tournament. To my considerable surprise, the winner was the simplest of all the programs submitted, TIT FOR TAT. TIT FOR TAT is merely the strategy of starting with cooperation, and thereafter doing what the other player did on the previous move.

I then circulated the results and solicited entries for a second round of the tournament. This time I received sixty-two entries from six countries. Most of the contestants were computer hobbyists, but there were also professors of evolutionary biology, physics, and computer science, as well as the five disciplines represented in the first round. As in the first round, some very elaborate programs were submitted. There were also a number of attempts to improve on TIT FOR TAT itself. TIT FOR TAT was again sent in by the winner of the first round, Anatol Rapoport of the University of Toronto. Again it won.

Something very interesting was happening here. I suspected that the properties that made TIT FOR TAT so successful in the tournaments would work in a world where *any* strategy was possible. If so, then cooperation based solely on reciprocity seemed possible. But I wanted to know the exact conditions that would be needed to foster cooperation on these terms. This led me to an evolutionary perspective: a consideration of how cooperation can emerge among egoists without central authority. The evolutionary perspective suggested three distinct questions. First, how can a potentially cooperative strategy get an initial foothold in an environment which is predominantly noncooperative? Second, what type of strategy can thrive in a variegated environment composed of other individuals using a wide diversity of more or less sophisticated strategies? Third, under what conditions can such a strategy, once fully established among a group of people, resist invasion by a less cooperative strategy?

The tournament results were published in the *Journal of Conflict Resolution* (Axelrod 1980*a* and 1980*b*), and are presented here in revised form in chapter 2. The theoretical results about initial viability, robustness, and stability were published in the *American Political Science Review* (Axelrod 1981). These findings provide the basis for chapter 3.

After thinking about the evolution of cooperation in a social context, I realized that the findings also had implications for biological evolution. So I collaborated with a biologist—William Hamilton—to develop the biological implications of these strategic ideas. This resulted in a paper published in *Science* (Axelrod and Hamilton 1981) which appears here in revised form as chapter 5. The paper has been awarded the Newcomb Cleveland Prize of the American Association for the Advancement of Science.

This gratifying response encouraged me to present these ideas in a form that would make them accessible not only to biologists and mathematically oriented social scientists but also to a broader audience interested in understanding the conditions that can foster cooperation among individuals, organizations, and nations. This in turn led me to see applications of the ideas in a great variety of concrete situations, and to appreciate how readily the results could be used to generate implications for private behavior and for public policy.

One point worth stressing at the outset is that this approach differs from that of sociobiology. Sociobiology is based on the assumption that important aspects of human behavior are guided by our genetic inheritance (e.g., E. O.

Wilson 1975). Perhaps so. But the present approach is *strategic* rather than *genetic.* It uses an evolutionary perspective because people are often in situations where effective strategies continue to be used and ineffective strategies are dropped. Sometimes the selection process is direct: a member of Congress who does not accomplish anything in interactions with colleagues will not long remain a member of Congress.

It is a pleasure to acknowledge the help received at various stages of this project from Jonathan Bendor, Robert Boyd, John Brehm, John Chamberlin, Joel Cohen, Lou Erste, John Ferejohn, Patty French, Bernard Grofman, Kenji Hayao, Douglas Hofstadter, Judy Jackson, Peter Katzenstein, William Keech, Martin Kessler, James March, Donald Markham, Richard Matland, John Meyer, Robert Mnookin, Larry Mohr, Lincoln Moses, Myra Oltsik, John Padgett, Jeff Pynnonen, Penelope Romlein, Amy Saldinger, Reinhart Selten, John David Sinclair, John T. Scholz, Serge Taylor, Robert Trivers, David Sloan Wilson, and especially Michael Cohen. I would also like to thank all the people whose entries made the tournaments possible. Their names are given in appendix A.

With gratitude I acknowledge the institutions that made this work possible: the Institute of Public Policy Studies of The University of Michigan, the Center for Advanced Study in the Behavioral Sciences, and the National Science Foundation under Grant SES-8023556.

FOREWORD TO THE NEW EDITION OF THE EVOLUTION OF COOPERATION

THIS IS A BOOK OF OPTIMISM. But it is a believable optimism, more satisfying than naïve, unrealistic hopes of pie in the sky (or rapture in the revolution).

To be believable, an optimism must first acknowledge fundamental reality, including the reality of human nature, but also the nature of all life. Life as we know it, and probably throughout the universe if there is life elsewhere, means Darwinian life. In a Darwinian world, that which survives survives, and the world becomes full of whatever qualities it takes to survive. As Darwinians, we start pessimistically by assuming deep selfishness at the level of natural selection, pitiless indifference to suffering, ruthless attention to individual success at the expense of others. And yet from such warped beginnings, something can come that is in effect, if not necessarily in intention, close to amicable brotherhood and sisterhood. This is the uplifting message of Robert Axelrod's remarkable book.

My own credentials for writing this foreword have been peripheral but recurrent. In the late 1970s, a few years after publishing my own first book, *The Selfish Gene,* which explained the pessimistic principles mentioned above, I received out of the blue a typescript from an American political scientist whom I didn't know: Robert Axelrod. It announced a "computer tournament" to play the game of Iterated Prisoner's Dilemma and invited me to compete. To be more precise—and the distinction is an important one for the very reason that the computer programs don't have conscious foresight—it invited me to submit a computer program that would do the competing. I'm afraid I didn't get around to sending in an entry. But I was hugely intrigued by the idea, and I did make one valuable, if rather passive, contribution to the enterprise at that stage. Axelrod was a professor of political science, and in my partisan way, I felt that he needed to collaborate with an evolutionary biologist. I wrote him an introduction to W. D. Hamilton, probably the most distinguished Darwinian of our generation, now sadly dead after an ill-fated expedition to the Congo jungle in 2000. In the 1970s, Hamilton was a colleague of Axelrod in a different department of the University of Michigan, but they didn't know each other. Upon receiving my letter, Axelrod immediately contacted Hamilton, and they collaborated on the paper that was the forerunner of this book and is abridged as Chapter 5. It had the same title as the book, was published in *Science* in 1981, and won the Newcomb-Cleveland prize of the American Association for the Advancement of Science.

The first American edition of *The Evolution of Cooperation* was published in 1984. I read it as soon as it appeared, with mounting excitement, and took to recommending it with evangelical zeal, to almost everyone I met. Every one of the Oxford undergraduates I tutored in the years following its publication was required to write an essay on Axelrod's book, and it was one of the essays they most enjoyed writing. But the book was not published in Britain, and in any case, the written word sadly has a limited constituency compared with television. So I was pleased when, in 1985, I was invited by Jeremy Taylor of the BBC to be the presenter of a *Horizon* program largely based upon Axelrod's work. We called the film *Nice Guys Finish First*. I had to speak my lines from such unaccustomed locations as a football pitch, a school in Britain's industrial midlands, a ruined medieval nunnery, a whooping cough vaccination clinic, and a replica of a First World War trench. *Nice Guys Finish First* appeared in the spring of 1986 and it enjoyed some critical success, although it was never shown in America—whether that is because of my unintelligible British accent I don't know. It also brought me temporary standing as a public partisan of "forgiving," "nonenvious," "nice guys"—a welcome relief, at least, from notoriety as the alleged high priest of selfishness, and salutary testimony to the power of title over content: My book had been *The Selfish Gene*, and I was regarded as an advocate of selfishness. My film was called *Nice Guys Finish First*, and I was hailed as Mr. Nice Guy. Neither accolade was borne out by the content of book or film. Nevertheless, in the weeks after *Nice Guys* was broadcast, I was lunched and consulted on niceness by industrialists and manufacturers. The chairman of Britain's leading chain of clothes shops gave me lunch in order to explain how nice his company was to its employees. A spokeswoman from a leading confectionery company also took me to lunch on a similar mission, in her case to explain that her company's dominant motivation in selling chocolate bars was not to make money but literally to spread sweetness and happiness among the population. Both, I fear, had slightly missed the point.

I was invited by the world's largest computer company to organize and supervise a whole day's game of strategy among their executives, whose purpose was to bond them together in amicable cooperation. They were divided into three teams—the reds, the blues, and the greens—and the game was a variant on the prisoner's dilemma game that is the central topic of this book. Unfortunately, the cooperative bonding that was the company's goal failed to materialize—spectacularly. As Robert Axelrod could have predicted, the fact that the game was known to be coming to an end at exactly 4 P.M. precipitated a massive defection by the reds against the blues immediately before the appointed hour. The bad feeling generated by this sudden break with the previous day-long goodwill was palpable at the post-mortem session that I conducted, and the executives had to have counseling before they could be persuaded to work together again.

In 1989, I acceded to Oxford University Press's request for a second edition of *The Selfish Gene.* It contains two chapters based upon the two books that most excited me during the intervening dozen years. It will come as no surprise that the first of these chapters was an exposition of Axelrod's work, again called *Nice Guys Finish First.* But I still felt that Axelrod's own book should be available in my own country. I took the initiative by approaching Penguin Books and was pleased that they accepted my recommendation to publish it, and they invited me to write a foreword to their British paperback edition. I am doubly pleased that Robert Axelrod himself has now invited me to update that foreword for this new edition of his book.

In the twenty-two years since *The Evolution of Cooperation* was first published, it is no exaggeration to say that it has spawned a whole new research industry. In 1988, Axelrod and a colleague, Douglas Dion, compiled an annotated bibliography of research publication more or less directly inspired by *The Evolution of Cooperation.* They listed more than 250 works up to that date under the following headings: "politics and law," "economics," "sociology and anthropology," "biological applications," "theory (including evolutionary theory)," "automata theory (computer science)," "new tournaments," and "miscellaneous." Axelrod and Dion collaborated on another paper published in *Science* (Volume 242, 1988, 1385–1390) with the title "The Further Evolution of Cooperation," summarizing the progress of the field in the four years since 1984. Since that review, nearly two decades have gone by and the growth of research fields inspired by this book has continued apace. The graph gives the numbers of *annual* citations of Robert Axelrod in the scientific

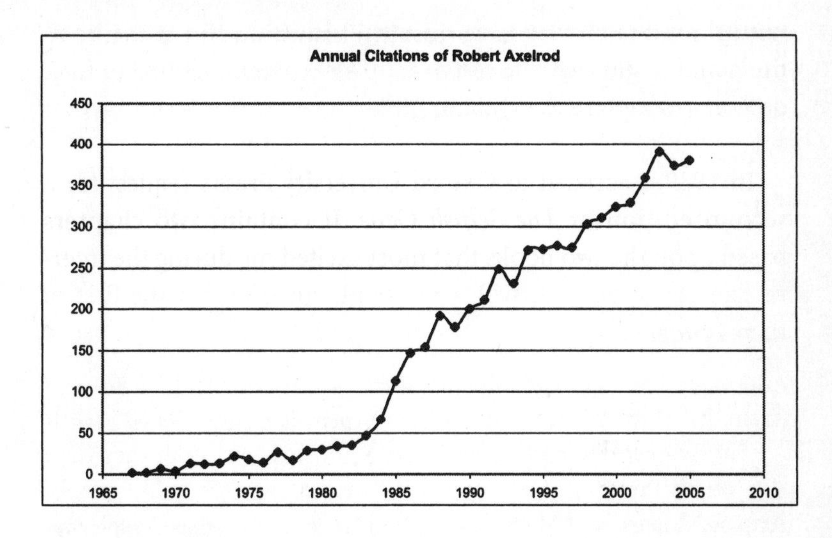

literature, and it clearly shows the impact that one influential book can have on the development of a field. Note the steep upturn of the graph after 1984, the publication date of *The Evolution of Cooperation.* Extensions of cooperation theory are found in books on prevention of war (Huth 1988), social evolution (Trivers 1985), cooperation among animals (Dugatkin 1997), human history (Wright 2000), evolutionary game theory (Gintis 2000), networks of trust and reciprocity that build social capital (Putnam 2000), microeconomics (Bowles 2004), science fiction (Anthony 1986), as well as books by Axelrod himself (1997 and 2001).

But in contemplating the welter of new research, the main impression I am left with is how little the basic conclusions of the book need to be changed. Ancient Mariner-like, I have continued over the years to press it upon students, colleagues, and passing acquaintances. I really do think that the planet would be a better place if everybody studied and understood it. The world's leaders should all be locked up with this book and not released until they have read it. This would be a pleasure to them and might save the rest of us. *The Evolution of Cooperation* deserves to replace the Gideon Bible.

RICHARD DAWKINS
Oxford, June 2006

References

Anthony, Piers. 1986. *Golem in the Gears*. New York: Ballantine Books.

Axelrod, Robert. 1997. *Complexity of Cooperation: Agent-Based Models of Competition and Cooperation*. Princeton, NJ: Princeton University Press.

Axelrod, Robert, and Michael D. Cohen. 2001. *Harnessing Complexity: Organizational Implications of a Scientific Frontier*. New York: Free Press.

Bowles, Samuel. 2004. *Microeconomics: Behavior, Institutions, and Evolution*. New York: Russell Sage Foundation and Princeton University Press.

Dugatkin, Lee Alan. 1997. *Cooperation Among Animals: An Evolutionary Perspective*. New York and Oxford: Oxford University Press.

Gintis, Herbert. 2000. *Game Theory Evolving: A Problem-Centered Introduction to Modeling Strategic Interaction*. Princeton, NJ: Princeton University Press.

Huth, Paul K. 1988. *Extended Deterrence and the Prevention of War*. New Haven, CT and London: Yale University Press.

Putnam, Robert D. 2000. *Bowling Alone: The Collapse and Revival of American Community*. New York: Simon & Schuster.

Trivers, Robert. 1985. *Social Evolution*. Menlo Park, CA: Benjamin/Cummings.

Wright, Robert. 2000. *Non-Zero: The Logic of Human Destiny*. New York: Pantheon.

I

Introduction

CHAPTER 1

The Problem of Cooperation

UNDER WHAT CONDITIONS will cooperation emerge in a world of egoists without central authority? This question has intrigued people for a long time. And for good reason. We all know that people are not angels, and that they tend to look after themselves and their own first. Yet we also know that cooperation does occur and that our civilization is based upon it. But, in situations where each individual has an incentive to be selfish, how can cooperation ever develop?

The answer each of us gives to this question has a fundamental effect on how we think and act in our social, political, and economic relations with others. And the answers that others give have a great effect on how ready they will be to cooperate with us.

The most famous answer was given over three hundred

years ago by Thomas Hobbes. It was pessimistic. He argued that before governments existed, the state of nature was dominated by the problem of selfish individuals who competed on such ruthless terms that life was "solitary, poor, nasty, brutish, and short" (Hobbes 1651/1962, p. 100). In his view, cooperation could not develop without a central authority, and consequently a strong government was necessary. Ever since, arguments about the proper scope of government have often focused on whether one could, or could not, expect cooperation to emerge in a particular domain if there were not an authority to police the situation.

Today nations interact without central authority. Therefore the requirements for the emergence of cooperation have relevance to many of the central issues of international politics. The most important problem is the security dilemma: nations often seek their own security through means which challenge the security of others. This problem arises in such areas as escalation of local conflicts and arms races. Related problems occur in international relations in the form of competition within alliances, tariff negotiations, and communal conflict in places like Cyprus.[1]

The Soviet invasion of Afghanistan in 1979 presented the United States with a typical dilemma of choice. If the United States continued business as usual, the Soviet Union might be encouraged to try other forms of noncooperative behavior later on. On the other hand, any substantial lessening of United States cooperation risked some form of retaliation, which could then set off counter-retaliation, setting up a pattern of mutual hostility that could be difficult to end. Much of the domestic debate about foreign policy is concerned with problems of just this type. And properly so, since these are hard choices.

In everyday life, we may ask ourselves how many times

we will invite acquaintances for dinner if they never invite us over in return. An executive in an organization does favors for another executive in order to get favors in exchange. A journalist who has received a leaked news story gives favorable coverage to the source in the hope that further leaks will be forthcoming. A business firm in an industry with only one other major company charges high prices with the expectation that the other firm will also maintain high prices—to their mutual advantage and at the expense of the consumer.

For me, a typical case of the emergence of cooperation is the development of patterns of behavior in a legislative body such as the United States Senate. Each senator has an incentive to appear effective to his or her constituents, even at the expense of conflicting with other senators who are trying to appear effective to *their* constituents. But this is hardly a situation of completely opposing interests, a zero-sum game. On the contrary, there are many opportunities for mutually rewarding activities by two senators. These mutually rewarding actions have led to the creation of an elaborate set of norms, or folkways, in the Senate. Among the most important of these is the norm of reciprocity—a folkway which involves helping out a colleague and getting repaid in kind. It includes vote trading but extends to so many types of mutually rewarding behavior that "it is not an exaggeration to say that reciprocity is a way of life in the Senate" (Matthews 1960, p. 100; see also Mayhew 1975).

Washington was not always like this. Early observers saw the members of the Washington community as quite unscrupulous, unreliable, and characterized by "falsehood, deceit, treachery" (Smith 1906, p. 190). In the 1980s the practice of reciprocity is well established. Even the significant changes in the Senate over the last two decades, tending toward more decentralization, more openness, and more equal distribution of power, have come without abating the folkway of reciprocity (Ornstein, Peabody, and Rhode 1977). As will be seen, it is *not* necessary to assume that senators are more honest, more generous, or more public-spirited than in earlier years to explain how cooperation based on reciprocity has emerged or proved stable. The emergence of cooperation can be explained as a consequence of individual senators pursuing their own interests.

The approach of this book is to investigate how individuals pursuing their own interests will act, followed by an analysis of what effects this will have for the system as a whole. Put another way, the approach is to make some assumptions about individual motives and then deduce consequences for the behavior of the entire system (Schelling 1978). The case of the U.S. Senate is a good example, but the same style of reasoning can be applied to other settings.

The object of this enterprise is to develop a theory of cooperation that can be used to discover what is necessary for cooperation to emerge. By understanding the conditions that allow it to emerge, appropriate actions can be taken to foster the development of cooperation in a specific setting.

The Cooperation Theory that is presented in this book is based upon an investigation of individuals who pursue their own self-interest without the aid of a central authority to force them to cooperate with each other. The reason for assuming self-interest is that it allows an examination of the difficult case in which cooperation is not completely based upon a concern for others or upon the welfare of the group as a whole. It must, however, be stressed that this assumption is actually much less restrictive than it appears.

If a sister is concerned for the welfare of her brother, the sister's self-interest can be thought of as including (among many other things) this concern for the welfare of her brother. But this does not necessarily eliminate all potential for conflict between sister and brother. Likewise a nation may act in part out of regard for the interests of its friends, but this regard does not mean that even friendly countries are always able to cooperate for their mutual benefit. So the assumption of self-interest is really just an assumption that concern for others does not completely solve the problem of when to cooperate with them and when not to.

A good example of the fundamental problem of cooperation is the case where two industrial nations have erected trade barriers to each other's exports. Because of the mutual advantages of free trade, both countries would be better off if these barriers were eliminated. But if either country were to unilaterally eliminate its barriers, it would find itself facing terms of trade that hurt its own economy. In fact, whatever one country does, the other country is better off retaining its own trade barriers. Therefore, the problem is that each country has an incentive to retain trade barriers, leading to a worse outcome than would have been possible had both countries cooperated with each other.

This basic problem occurs when the pursuit of self-interest by each leads to a poor outcome for all. To make headway in understanding the vast array of specific situations which have this property, a way is needed to represent what is common to these situations without becoming bogged down in the details unique to each. Fortunately, there is such a representation available: the famous *Prisoner's Dilemma* game.[2]

In the Prisoner's Dilemma game, there are two players. Each has two choices, namely cooperate or defect. Each must make the choice without knowing what the other will do. No matter what the other does, defection yields a higher payoff than cooperation. The dilemma is that if both defect, both do worse than if both had cooperated. This simple game will provide the basis for the entire analysis used in this book.

The way the game works is shown in figure 1. One player chooses a row, either cooperating or defecting. The other player simultaneously chooses a column, either cooperating or defecting. Together, these choices result in one of the four possible outcomes shown in that matrix. If both players cooperate, both do fairly well. Both get *R*, the *reward for mutual cooperation.* In the concrete illustration of figure 1 the reward is 3 points. This number might, for example, be a payoff in dollars that each player gets for that outcome. If one player cooperates but the other defects, the defecting player gets the *temptation to defect,* while the cooperating player gets the *sucker's payoff.* In the example, these are 5 points and 0 points respectively. If both defect, both get 1 point, the *punishment for mutual defection.*

What should you do in such a game? Suppose you are the row player, and you think the column player will coop-

FIGURE 1
The Prisoner's Dilemma

		Column Player	
		Cooperate	*Defect*
Row Player	Cooperate	R=3, R=3 Reward for mutual cooperation	S=0, T=5 Sucker's payoff, and temptation to defect
	Defect	T=5, S=0 Temptation to defect and sucker's payoff	P=1, P=1 Punishment for mutual defection

NOTE: The payoffs to the row chooser are listed first.

erate. This means that you will get one of the two outcomes in the first column of figure 1. You have a choice. You can cooperate as well, getting the 3 points of the reward for mutual cooperation. Or you can defect, getting the 5 points of the temptation payoff. So it pays to defect if you think the other player will cooperate. But now suppose that you think the other player will defect. Now you are in the second column of figure 1, and you have a choice between cooperating, which would make you a sucker and give you 0 points, and defecting, which would result in, mutual punishment giving you 1 point. So it pays to defect if you think the other player will defect. This means that it is better to defect if you think the other player will cooperate, *and* it is better to defect if you think the other player will defect. So no matter what the other player does, it pays for you to defect.

So far, so good. But the same logic holds for the other player too. Therefore, the other player should defect no matter what you are expected to do. So you should both defect. But then you both get 1 point which is worse than the 3 points of the reward that you both could have gotten had you both cooperated. Individual rationality leads to a worse outcome for both than is possible. Hence the dilemma.

The Prisoner's Dilemma is simply an abstract formulation of some very common and very interesting situations in which what is best for each person individually leads to mutual defection, whereas everyone would have been better off with mutual cooperation. The definition of Prisoner's Dilemma requires that several relationships hold among the four different potential outcomes. The first relationship specifies the order of the four payoffs. The best a player can do is get T, the temptation to defect when the other player cooperates. The worst a player can do is get *S,* the sucker's payoff for cooperating while the other player defects. In ordering the other two outcomes, *R,* the reward for mutual cooperation, is assumed to be better than *P,* the punishment for mutual defection. This leads to a preference ranking of the four payoffs from best to worst as *T, R, P,* and *S.*

The second part of the definition of the Prisoner's Dilemma is that the players cannot get out of their dilemma by taking turns exploiting each other. This assumption means that an even chance of exploitation and being exploited is not as good an outcome for a player as mutual cooperation. It is therefore assumed that the reward for mutual cooperation is greater than the average of the temptation and the sucker's payoff. This assumption, together with the rank ordering of the four payoffs, defines the Prisoner's Dilemma.

Thus two egoists playing the game *once* will both choose their dominant choice, defection, and each will get less than they both could have gotten if they had cooperated. If the game is played a known finite number of times, the players still have no incentive to cooperate. This is certainly true on the last move since there is no future to influence. On the next-to-last move neither player will have an incentive to cooperate since they can both anticipate a defection by the other player on the very last move. Such a line of reasoning implies that the game will unravel all the way back to mutual defection on the first move of any sequence of plays that is of known finite length (Luce and Raiffa 1957, pp. 94–102). This reasoning does not apply if the players will interact an indefinite number of times. And in most realistic settings, the players cannot be sure when the last interaction between them will take place. As will be shown later, with an indefinite number of interactions, cooperation can emerge. The issue then becomes the discovery of the precise conditions that are necessary and sufficient for cooperation to emerge.

In this book I will examine interactions between just two players at a time. A single player may be interacting with many others, but the player is assumed to be interacting with them one at a time.[3] The player is also assumed to recognize another player and to remember how the two of them have interacted so far. This ability to recognize and remember allows the history of the particular interaction to be taken into account by a player's strategy.

A variety of ways to resolve the Prisoner's Dilemma have been developed. Each involves allowing some additional activity that alters the strategic interaction in such a way as to fundamentally change the nature of the problem. The original problem remains, however, because there are many situations in which these remedies are not available. Therefore, the problem will be considered in its fundamental form, without these alterations.

1. There is no mechanism available to the players to make enforceable threats or commitments (Schelling 1960). Since the players cannot commit themselves to a particular strategy, each must take into account all possible strategies that might be used by the other player. Moreover the players have all possible strategies available to themselves.

2. There is no way to be sure what the other player will do on a given move. This eliminates the possibility of metagame analysis (Howard 1971) which allows such options as "make the same choice as the other is about to make." It also eliminates the possibility of reliable reputations such as might be based on watching the other player interact with

third parties. Thus the only information available to the players about each other is the history of their interaction so far.

3. There is no way to eliminate the other player or run away from the interaction. Therefore each player retains the ability to cooperate or defect on each move.

4. There is no way to change the other player's payoffs. The payoffs already include whatever consideration each player has for the interests of the other (Taylor 1976, pp. 69–73).

Under these conditions, words not backed by actions are so cheap as to be meaningless. The players can communicate with each other only through the sequence of their own behavior. This is the problem of the Prisoner's Dilemma in its fundamental form.

What makes it possible for cooperation to emerge is the fact that the players might meet again. This possibility means that the choices made today not only determine the outcome of this move, but can also influence the later choices of the players. The future can therefore cast a shadow back upon the present and thereby affect the current strategic situation.

But the future is less important than the present—for two reasons. The first is that players tend to value payoffs less as the time of their obtainment recedes into the future. The second is that there is always some chance that the players will not meet again. An ongoing relationship may end when one or the other player moves away, changes jobs, dies, or goes bankrupt.

For these reasons, the payoff of the next move always counts less than the payoff of the current move. A natural way to take this into account is to cumulate payoffs over time in such a way that the next move is worth some fraction of the current move (Shubik 1970). The *weight* (or importance) of the next move relative to the current move will be called w. It represents the degree to which the payoff of each move is discounted relative to the previous move, and is therefore a *discount parameter.*

The discount parameter can be used to determine the payoff for a whole sequence. To take a simple example, suppose that each move is only half as important as the previous move, making $w = 1/2$. Then a whole string of mutual defections worth one point each move would have a value of 1 on the first move, ½ on the second move, ¼ on the third move, and so on. The cumulative value of the sequence would be $1 + 1/2 + 1/4 + 1/8 \ldots$ which would sum to exactly 2. In general, getting one point on each move would be worth $1 + w + w^2 + w^3 \ldots$ A very useful fact is that the sum of this infinite series for any w greater than zero and less than one is simply $1/(1-w)$. To take another case, if each move is worth 90 percent of the previous move, a string of 1's would be worth ten points because $1/(1-w) = 1/(1-.9) = 1/.1 = 10$. Similarly, with w still equal to .9, a string of 3 point mutual rewards would be worth three times this, or 30 points.

Now consider an example of two players interacting. Suppose one player is following the policy of always defecting (ALL D), and the other player is following the policy of TIT FOR TAT. TIT FOR TAT is the policy of cooperating on the first move and then doing whatever the other player did on the previous move. This policy means that TIT FOR TAT will defect once after each defection of the other player. When the other player is using TIT FOR TAT, a player who always defects will get T on the first move, and P on all subsequent moves. The *value* (or *score*) to someone using ALL D when playing with someone using TIT FOR TAT is thus the sum of T for the first move, wP for the second move, w^2P for the third move, and so on.[4]

Both ALL D and TIT FOR TAT are strategies. In general, a *strategy* (or *decision rule*) is a specification of what to do in any situation that might arise. The situation itself depends upon the history of the game so far. Therefore, a strategy might cooperate after some patterns of interaction and defect after others. Moreover, a strategy may use probabilities, as in the example of a rule which is entirely random with equal probabilities of cooperation and defection on each move. A strategy can also be quite sophisticated in its use of the pattern of outcomes in the game so far to determine what to do next. An example is one which, on each move, models the behavior of the other player using a complex procedure (such as a Markov process), and then uses a fancy method of statistical inference (such as Bayesian analysis) to select what seems the best choice for the long run. Or it may be some intricate combination of other strategies.

The first question you are tempted to ask is, "What is the best strategy?" In other words, what strategy will yield a player the highest possible score? This is a good question, but as will be shown later, no best rule exists independently of the strategy being used by the other player. In this sense, the iterated Prisoner's Dilemma is completely different from a game like chess. A chess master can safely use the assumption that the other player will make the most feared move. This assumption provides a basis for planning in a game like chess, where the interests of the players are completely antagonistic. But the situations represented by the Prisoner's Dilemma game are quite different. The interests of the players are not in total conflict. Both players can do well by getting the reward, *R,* for mutual cooperation or both can do poorly by getting the punishment, *P,* for mutual defection. Using the assumption that the other player will always make the move you fear most will lead you to expect that the other will never cooperate, which in turn will lead you to defect, causing unending punishment. So unlike chess, in the Prisoner's Dilemma it is not safe to assume that the other player is out to get you.

In fact, in the Prisoner's Dilemma, the strategy that works best depends directly on what strategy the other player is using and, in particular, on whether this strategy leaves room for the development of mutual cooperation. This principle is based on the weight of the next move relative to the current move being sufficiently large to make the future important. In other words, the discount parameter, *w,* must be large enough to make the future loom large in the calculation of total payoffs. After all, if you are unlikely to meet the other person again, or if you care little about future payoffs, then you might as well defect now and not worry about the consequences for the future.

This leads to the first formal proposition. It is the sad news that if the future is important, there is no one best strategy.

Proposition 1. If the discount parameter, *w,* is sufficiently high, there is no best strategy independent of the strategy used by the other player.

The proof itself is not hard. Suppose that the other player is using ALL D, the strategy of always defecting. If the other player will never cooperate, the best you can do is always to defect yourself. Now suppose, on the other hand, that the other player is using a strategy of "permanent retaliation." This is the strategy of cooperating until you defect and then always defecting after that. In that case, your best strategy is never to defect, provided that the temptation to defect on the first move will eventually be more than compensated for by the long-term disadvantage of getting nothing but the punishment, *P*, rather than the reward, *R*, on future moves. This will be true whenever the discount parameter, *w*, is sufficiently great.[5] Thus, whether or not you should cooperate, even on the first move, depends on the strategy being used by the other player. Therefore, if *w* is sufficiently large, there is no one best strategy.

In the case of a legislature such as the U.S. Senate, this proposition says that if there is a large enough chance that a member of the legislature will interact *again* with another member, there is no one best strategy to use independently of the strategy being used by the other person. It would be best to cooperate with someone who will reciprocate that cooperation in the future, but not with someone whose future behavior will not be very much affected by this interaction (see, for example, Hinckley 1972). The very possibility of achieving stable mutual cooperation depends upon there being a good chance of a continuing interaction, as measured by the magnitude of *w*. As it happens, in the case of Congress, the chance of two members having a continuing interaction has increased dramatically as the biennial turnover rates have fallen from about 40 percent in the first forty years of the republic to about 20 percent or less in recent years (Young 1966, pp. 87–90; Polsby 1968; Jones 1977, p. 154; Patterson 1978, pp. 143–44).

However, saying that a continuing chance of interaction is necessary for the development of cooperation is not the same as saying that it is sufficient. The demonstration that there is not a single best strategy leaves open the question of what patterns of behavior can be expected to emerge when there actually is a sufficiently high probability of continuing interaction between two individuals.

Before going on to study the behavior that can be expected to emerge, it is a good idea to take a closer look at which features of reality the Prisoner's Dilemma framework is, and is not, able to encompass. Fortunately, the very simplicity of the framework makes it possible to avoid many restrictive assumptions that would otherwise limit the analysis:

1. The payoffs of the players need not be comparable at all. For example, a journalist might get rewarded with another inside story, while the cooperating bureaucrat might be rewarded with a chance to have a policy argument presented in a favorable light.

2. The payoffs certainly do not have to be symmetric. It is a convenience to think of the interaction as exactly equivalent from the perspective of the two players, but this is not necessary. One does not have to assume, for example, that the reward for mutual cooperation, or any of the other three payoff parameters, have the same magnitude for both players. As mentioned earlier, one does not even have to assume that they are measured in comparable units. The only thing that has to be assumed is that, for each player, the four payoffs are ordered as required for the definition of the Prisoner's Dilemma.

3. The payoffs of a player do not have to be measured on an absolute scale. They need only be measured relative to each other.[6]

4. Cooperation need not be considered desirable from the point of view of the rest of the world. There are times when one wants to retard, rather than foster, cooperation between players. Collusive business practices are good for

the businesses involved but not so good for the rest of society. In fact, most forms of corruption are welcome instances of cooperation for the participants but are unwelcome to everyone else. So, on occasion, the theory will be used in reverse to show how to prevent, rather than to promote, cooperation.

5. There is no need to assume that the players are rational. They need not be trying to maximize their rewards. Their strategies may simply reflect standard operating procedures, rules of thumb, instincts, habits, or imitation (Simon 1955; Cyert and March 1963).

6. The actions that players take are not necessarily even conscious choices. A person who sometimes returns a favor, and sometimes does not, may not think about what strategy is being used. There is no need to assume deliberate choice at all.[7]

The framework is broad enough to encompass not only people but also nations and bacteria. Nations certainly take actions which can be interpreted as choices in a Prisoner's Dilemma—as in the raising or lowering of tariffs. It is not necessary to assume that such actions are rational or are the outcome of a unified actor pursuing a single goal. On the contrary, they might well be the result of an incredibly complex bureaucratic politics involving complicated information processing and shifting political coalitions (Allison 1971).

Likewise, at the other extreme, an organism does not need a brain to play a game. Bacteria, for example, are highly responsive to selected aspects of their chemical environment. They can therefore respond differentially to what other organisms are doing, and these conditional strategies of behavior can be inherited. Moreover, the behavior of a bacterium can affect the fitness of other organisms around it, just as the behavior of other organisms can affect the fitness of a bacterium. But biological applications are best saved for chapter 5.

For now the main interest will be in people and organizations. Therefore, it is good to know that for the sake of generality, it is not necessary to assume very much about how deliberate and insightful people are. Nor is it necessary to assume, as the sociobiologists do, that important aspects of human behavior are guided by one's genes. The approach here is strategic rather than genetic.

Of course, the abstract formulation of the problem of cooperation as a Prisoner's Dilemma puts aside many vital features that make any actual interaction unique. Examples of what is left out by this formal abstraction include the possibility of verbal communication, the direct influence of third parties, the problems of implementing a choice, and the uncertainty about what the other player actually did on the preceding move. In chapter 8 some of these complicating factors are added to the basic model. It is clear that the list of potentially relevant factors that have been left out could be extended almost indefinitely. Certainly, no intelligent person should make an important choice without trying to take such complicating factors into account. The value of an analysis without them is that it can help to clarify some of the subtle features of the interaction—features which might otherwise be lost in the maze of complexities of the highly particular circumstances in which choice must actually be made. It is the very complexity of reality which makes the analysis of an abstract interaction so helpful as an aid to understanding.

The next chapter explores the emergence of cooperation through a study of what is a good strategy to employ if confronted with an iterated Prisoner's Dilemma. This exploration has been done in a novel way, with a computer tournament. Professional game theorists were invited to submit their favorite strategy, and each of these decision rules was paired off with each of the others to see which would do best overall. Amazingly enough, the winner was the simplest of all strategies submitted. This was TIT FOR TAT, the strategy which cooperates on the first move and then does whatever the other player did on the previous move. A second round of the tournament was conducted in which many more entries were submitted by amateurs and professionals alike, all of whom were aware of the results of the first round. The result was another victory for TIT FOR TAT! The analysis of the data from these tournaments reveals four properties which tend to make a decision rule successful: avoidance of unnecessary conflict by cooperating as long as the other player does, provocability in the face of an uncalled for defection by the other, forgiveness after responding to a provocation, and clarity of behavior so that the other player can adapt to your pattern of action.

These results from the tournaments demonstrate that under suitable conditions, cooperation can indeed emerge in a world of egoists without central authority. To see just how widely these results apply, a theoretical approach is taken in chapter 3. A series of propositions are proved that not only demonstrate the requirements for the emergence of cooperation but also provide the chronological story of the evolution of cooperation. Here is the argument in a nutshell. The evolution of cooperation requires that individuals have a sufficiently large chance to meet again so that they have a stake in their future interaction. If this is true, cooperation can evolve in three stages.

1. The beginning of the story is that cooperation can get started even in a world of unconditional defection. The development *cannot* take place if it is tried only by scattered individuals who have virtually no chance to interact with each other. However, cooperation can evolve from small clusters of individuals who base their cooperation on reciprocity and have even a small proportion of their interactions with each other.

2. The middle of the story is that a strategy based on reciprocity can thrive in a world where many different kinds of strategies are being tried.

3. The end of the story is that cooperation, once established on the basis of reciprocity, can protect itself from invasion by less cooperative strategies. Thus, the gear wheels of social evolution have a ratchet.

Chapters 4 and 5 take concrete settings to demonstrate just how widely these results apply. Chapter 4 is devoted to the fascinating case of the "live and let live" system which emerged during the trench warfare of World War I. In the midst of this bitter conflict, the front-line soldiers often refrained from shooting to kill—provided their restraint was reciprocated by the soldiers on the other side. What made this mutual restraint possible was the static nature of trench warfare, where the same small units faced each other for extended periods of time. The soldiers of these opposing small units actually violated orders from their own high commands in order to achieve tacit cooperation with each other. A detailed look at this case shows that when the conditions are present for the emergence of cooperation, cooperation can get started and prove stable in situations which otherwise appear extraordinarily unpromising. In particular, the "live and let live" system demonstrates that friendship is hardly necessary for the development of cooperation. Under suitable conditions, cooperation based upon reciprocity can develop even between antagonists.

Chapter 5, written with evolutionary biologist William D. Hamilton, demonstrates that cooperation can emerge even without foresight. This is done by showing that Cooperation Theory can account for the patterns of behavior found in a wide range of biological systems, from bacteria to birds. Cooperation in biological systems can occur even when the participants are not related, and even when they are unable to appreciate the consequences of their own behavior. What makes this possible are the evolutionary mechanisms of genetics and survival of the fittest. An individual able to achieve a beneficial response from another is more likely to have offspring that survive and that continue the pattern of behavior which elicited beneficial responses from others. Thus, under suitable conditions, cooperation based upon reciprocity proves stable in the biological world. Potential applications are spelled out for specific aspects of territoriality, mating, and disease. The conclusion is that Darwin's emphasis on individual advantage can, in fact, account for the presence of cooperation between individuals of the same or even different species. As long as the proper conditions are present, cooperation can get started, thrive, and prove stable.

While foresight is not necessary for the evolution of cooperation, it can certainly be helpful. Therefore chapters 6 and 7 are devoted to offering advice to participants and reformers, respectively. Chapter 6 spells out the implications of Cooperation Theory for anyone who is in a Prisoner's Dilemma. From the participant's point of view, the object is to do as well as possible, regardless of how well the other player does. Based upon the tournament results and the formal propositions, four simple suggestions are offered for individual choice: do not be envious of the other player's success; do not be the first to defect; reciprocate both cooperation and defection; and do not be too clever.

Understanding the perspective of a participant can also serve as the foundation for seeing what can be done to make it easier for cooperation to develop among egoists. Thus, chapter 7 takes the Olympian perspective of a reformer who wants to alter the very terms of the interactions so as to promote the emergence of cooperation. A wide variety of methods are considered, such as making the interactions between the players more durable and frequent, teaching the participants to care about each other, and teaching them to understand the value of reciprocity. This reformer's perspective provides insights into a wide variety of topics, from the strength of bureaucracy to the difficulties of Gypsies, and from the morality of TIT FOR TAT to the art of writing treaties.

Chapter 8 extends the implications of Cooperation Theory into new domains. It shows how different kinds of social structure affect the way cooperation can develop. For example, people often relate to each other in ways that are influenced by observable features, such as sex, age, skin color, and style of dress. These cues can lead to social structures based on stereotyping and status hierarchies. As another example of social structure, the role of reputation is considered. The struggle to establish and maintain one's reputation can be a major feature of intense conflicts. For example, the American government's escalation of the war in Vietnam in 1965 was mainly due to its desire to deter other challenges to its interests by maintaining its reputation on the world stage. This chapter also considers a government's concern for maintaining its reputation with its own citizens. To be effective, a government cannot enforce any standards it chooses but must elicit compliance from a majority of the governed. To do this requires setting the rules so that most of the governed find it profitable to obey most of the time. The implications of this approach are fundamental to the operation of authority, and are illustrated by the regulation of industrial pollution and the supervision of divorce settlements.

By the final chapter, the discussion has developed from the study of the emergence of cooperation among egoists without central authority to an analysis of what happens when people actually *do* care about each other and what happens when there *is* central authority. But the basic approach is always the same: seeing how individuals operate in their own interest reveals what happens to the whole group. This approach allows more than the understanding of the perspective of a single player. It also provides an appreciation of what it takes to promote the stability of mutual cooperation in a given setting. The most promising finding is that if the facts of Cooperation Theory are known by participants with foresight, the evolution of cooperation can be speeded up.

II

The Emergence of Cooperation

CHAPTER 2

The Success of TIT FOR TAT in Computer Tournaments

SINCE the Prisoner's Dilemma is so common in everything from personal relations to international relations, it would be useful to know how best to act when in this type of setting. However, the proposition of the previous chapter demonstrates that there is no one best strategy to use. What is best depends in part on what the other player is likely to be doing. Further, what the other is likely to be doing may well depend on what the player expects *you* to do.

To get out of this tangle, help can be sought by combing the research already done concerning the Prisoner's Dilemma for useful advice. Fortunately, a great deal of research has been done in this area.

Psychologists using experimental subjects have found

that, in the iterated Prisoner's Dilemma, the amount of cooperation attained—and the specific pattern for attaining it—depend on a wide variety of factors relating to the context of the game, the attributes of the individual players, and the relationship between the players. Since behavior in the game reflects so many important factors about people, it has become a standard way to explore questions in social psychology, from the effects of westernization in Central Africa (Bethlehem 1975) to the existence (or nonexistence) of aggression in career-oriented women (Baefsky and Berger 1974), and to the differential consequences of abstract versus concrete thinking styles (Nydegger 1974). In the last fifteen years, there have been hundreds of articles on the Prisoner's Dilemma cited in *Psychological Abstracts.* The iterated Prisoner's Dilemma has become the *E. coli* of social psychology.

Just as important as its use as an experimental test bed is the use of the Prisoner's Dilemma as the conceptual foundation for models of important social processes. Richardson's model of the arms race is based on an interaction which is essentially a Prisoner's Dilemma, played once a year with the budgets of the competing nations (Richardson 1960; Zinnes 1976, pp. 330–40). Oligopolistic competition can also be modeled as a Prisoner's Dilemma (Samuelson 1973, pp. 503–5). The ubiquitous problems of collective action to produce a collective good are analyzable as Prisoner's Dilemmas with many players (G. Hardin 1982). Even vote trading has been modeled as a Prisoner's Dilemma (Riker and Brams 1973). In fact, many of the best-developed models of important political, social, and economic processes have the Prisoner's Dilemma as their foundation.

There is yet a third literature about the Prisoner's Dilemma. This literature goes beyond the empirical questions of the laboratory or the real world, and instead uses the abstract game to analyze the features of some fundamental strategic issues, such as the meaning of rationality (Luce and Raiffa 1957), choices which affect other people (Schelling 1973), and cooperation without enforcement (Taylor 1976).

Unfortunately, none of these three literatures on the Prisoner's Dilemma reveals very much about how to play the game well. The experimental literature is not much help, because virtually all of it is based on analyzing the choices made by players who are seeing the formal game for the first time. Their appreciation of the strategic subtleties is bound to be restricted. Although the experimental subjects may have plenty of experience with everyday occurrences of the Prisoner's Dilemma, their ability to call on this experience in a formal setting may be limited. The choices of experienced economic and political elites in natural settings are studied in some of the applied literature of Prisoner's Dilemma, but the evidence is of limited help because of the relatively slow pace of most high-level interactions and the difficulty of controlling for changing circumstances. All together, no more than a few dozen choices have been identified and analyzed this way. Finally, the abstract literature of strategic interaction usually studies variants of the iterated Prisoner's Dilemma designed to eliminate the dilemma itself by introducing changes in the game, such as allowing interdependent choices (Howard 1966; Rapoport 1967), or putting a tax on defection (Tideman and Tullock 1976; Clarke 1980).

To learn more about how to choose effectively in an iterated Prisoner's Dilemma, a new approach is needed. Such an approach would have to draw on people who have a rich understanding of the strategic possibilities inherent in a non-zero-sum setting, a situation in which the interests of the participants partially coincide and partially conflict. Two important facts about non-zero-sum settings would have to be taken into account. First, the proposition of the previous chapter demonstrates that what is effective depends not only upon the characteristics of a particular strategy, but also upon the nature of the other strategies with which it must interact. The second point follows directly from the first. An effective strategy must be able at any point to take into account the history of the interaction as it has developed so far.

A computer tournament for the study of effective choice in the iterated Prisoner's Dilemma meets these needs. In a computer tournament, each entrant writes a program that embodies a rule to select the cooperative or noncooperative choice on each move. The program has available to it the history of the game so far, and may use this history in making a choice. If the participants are recruited primarily from those who are familiar with the Prisoner's Dilemma, the entrants can be assured that their decision rule will be facing rules of other informed entrants. Such recruitment would also guarantee that the state of the art is represented in the tournament.

Wanting to find out what would happen, I invited professional game theorists to send in entries to just such a computer tournament. It was structured as a round robin, meaning that each entry was paired with each other entry. As announced in the rules of the tournament, each entry was also paired with its own twin and with RANDOM, a program that randomly cooperates and defects with equal probability. Each game consisted of exactly two hundred moves.[1] The payoff matrix for each move was the familiar one described in chapter 1. It awarded both players 3 points for mutual cooperation, and 1 point for mutual defection. If one player defected while the other player cooperated, the defecting player received 5 points and the cooperating player received 0 points.

No entry was disqualified for exceeding the allotted time. In fact, the entire round robin tournament was run five times to get a more stable estimate of the scores for each pair of players. In all, there were 120,000 moves, making for 240,000 separate choices.

The fourteen submitted entries came from five disciplines: psychology, economics, political science, mathematics, and sociology. Appendix A lists the names and affiliations of the people who submitted these entries, and it gives the rank and score of their entries.

One remarkable aspect of the tournament was that it allowed people from different disciplines to interact with each other in a common format and language. Most of the entrants were recruited from those who had published articles on game theory in general or the Prisoner's Dilemma in particular.

TIT FOR TAT, submitted by Professor Anatol Rapoport of the University of Toronto, won the tournament. This was the simplest of all submitted programs and it turned out to be the best!

TIT FOR TAT, of course, starts with a cooperative choice, and thereafter does what the other player did on the previous move. This decision rule is probably the most widely known and most discussed rule for playing the Prisoner's Dilemma. It is easily understood and easily programmed. It is known to elicit a good degree of cooperation when played with humans (Oskamp 1971; W. Wilson 1971). As an entry in a computer tournament, it has the desirable properties that it is not very exploitable and that it does well with its own twin. It has the disadvantage that it is too generous with the RANDOM rule, which was known by the participants to be entered in the tournament.

In addition, TIT FOR TAT was known to be a powerful competitor. In a preliminary tournament, TIT FOR TAT scored second place; and in a variant of that preliminary tournament, TIT FOR TAT won first place. All of these facts were known to most of the people designing programs for the Computer Prisoner's Dilemma Tournament, because they were sent copies of a description of the preliminary tournament. Not surprisingly, many of them used the TIT FOR TAT principle and tried to improve upon it.

The striking fact is that *none* of the more complex programs submitted was able to perform as well as the original, simple TIT FOR TAT.

This result contrasts with computer chess tournaments, where complexity is obviously needed. For example, in the Second World Computer Chess Championships, the least complex program came in last (Jennings 1978). It was submitted by Johann Joss of the Eidgenossishe Technische Hochschule of Zurich, Switzerland, who also submitted an entry to the Computer Prisoner's Dilemma Tournament. His entry to the Prisoner's Dilemma Tournament was a small modification of TIT FOR TAT. But his modification, like the others, just lowered the performance of the decision rule.

Analysis of the results showed that neither the discipline of the author, the brevity of the program—nor its *length*—accounts for a rule's relative success. What does?

Before answering this question, a remark on the interpretation of numerical scores is in order. In a game of 200 moves, a useful benchmark for very good performance is

600 points, which is equivalent to the score attained by a player when both sides always cooperate with each other. A useful benchmark for very poor performance is 200 points, which is equivalent to the score attained by a player when both sides never cooperate with each other. Most scores range between 200 and 600 points, although scores from 0 to 1000 points are possible. The winner, TIT FOR TAT, averaged 504 points per game.

Surprisingly, there is a single property which distinguishes the relatively high-scoring entries from the relatively low-scoring entries. This is the property of being *nice,* which is to say never being the first to defect. (For the sake of analyzing this tournament, the definition of a nice rule will be relaxed to include rules which will not be the first to defect before the last few moves, say before move 199.)

Each of the eight top-ranking entries (or rules) is nice. None of the other entries is. There is even a substantial gap in the score between the nice entries and the others. The nice entries received tournament averages between 472 and 504, while the best of the entries that were not nice received only 401 points. Thus, not being the first to defect, at least until virtually the end of the game, was a property which, all by itself, separated the more successful rules from the less successful rules in this Computer Prisoner's Dilemma Tournament.

Each of the nice rules got about 600 points with each of the other seven nice rules and with its own twin. This is because when two nice rules play, they are sure to cooperate with each other until virtually the end of the game. Actually the minor variations in end-game tactics did not account for much variation in the scores.

Since the nice rules all got within a few points of 600

with each other, the thing that distinguished the relative rankings among the nice rules was their scores with the rules which are not nice. This much is obvious. What is not obvious is that the relative ranking of the eight top rules was largely determined by just two of the other seven rules. These two rules are *kingmakers* because they do not do very well for themselves, but they largely determine the rankings among the top contenders.

The most important kingmaker was based on an "outcome maximization" principle originally developed as a possible interpretation of what human subjects do in the Prisoner's Dilemma laboratory experiments (Downing 1975). This rule, called DOWNING, is a particularly interesting rule in its own right. It is well worth studying as an example of a decision rule which is based upon a quite sophisticated idea. Unlike most of the others, its logic is not just a variant of TIT FOR TAT. Instead it is based on a deliberate attempt to understand the other player and then to make the choice that will yield the best long-term score based upon this understanding. The idea is that if the other player does not seem responsive to what DOWNING is doing, DOWNING will try to get away with whatever it can by defecting. On the other hand, if the other player does seem responsive, DOWNING will cooperate. To judge the other's responsiveness, DOWNING estimates the probability that the other player cooperates after it (DOWNING) cooperates, and also the probability that the other player cooperates after DOWNING defects. For each move, it updates its estimate of these two conditional probabilities and then selects the choice which will maximize its own long-term payoff under the assumption that it has correctly modeled the other player. If the two conditional probabilities have similar values, DOWNING determines that it pays to defect, since the other player seems to be doing the same thing whether DOWNING cooperates or not. Conversely, if the other player tends to cooperate after a cooperation but not after a defection by DOWNING, then the other player seems responsive, and DOWNING will calculate that the best thing to do with a responsive player is to cooperate. Under certain circumstances, DOWNING will even determine that the best strategy is to alternate cooperation and defection.

At the start of a game, DOWNING does not know the values of these conditional probabilities for the other players. It assumes that they are both .5, but gives no weight to this estimate when information actually does come in during the play of the game.

This is a fairly sophisticated decision rule, but its implementation does have one flaw. By initially assuming that the other player is unresponsive, DOWNING is doomed to defect on the first two moves. These first two defections led many other rules to punish DOWNING, so things usually got off to a bad start. But this is precisely why DOWNING served so well as a kingmaker. First-ranking TIT FOR TAT and second-ranking TIDEMAN AND CHIERUZZI both reacted in such a way that DOWNING learned to expect that defection does not pay but that cooperation does. All of the other nice rules went downhill with DOWNING.

The nice rules did well in the tournament largely because they did so well with each other, and because there were enough of them to raise substantially each other's average score. As long as the other player did not defect, each of the nice rules was certain to continue cooperating until virtually the end of the game. But what happened if there was a defection? Different rules responded quite differently, and their response was important in determining their overall success. A key concept in this regard is the forgiveness of a decision rule. *Forgiveness* of a rule can be informally described as its propensity to cooperate in the moves after the other player has defected.[2]

Of all the nice rules, the one that scored lowest was also the one that was least forgiving. This is FRIEDMAN, a totally unforgiving rule that employs permanent retaliation. It is never the first to defect, but once the other defects even once, FRIEDMAN defects from then on. In contrast, the winner, TIT FOR TAT, is unforgiving for one move, but thereafter is totally forgiving of that defection. After one punishment, it lets bygones be bygones.

One of the main reasons why the rules that are not nice did not do well in the tournament is that most of the rules in the tournament were not very forgiving. A concrete illustration will help. Consider the case of JOSS, a sneaky rule that tries to get away with an occasional defection. This decision rule is a variation of TIT FOR TAT. Like TIT FOR TAT, it always defects immediately after the other player defects. But instead of always cooperating after the other player cooperates, 10 percent of the time it defects after the other player cooperates. Thus it tries to sneak in an occasional exploitation of the other player.

This decision rule seems like a fairly small variation of TIT FOR TAT, but in fact its overall performance was much worse, and it is interesting to see exactly why. Table 1 shows the move-by-move history of a game between JOSS and TIT FOR TAT. At first both players cooperated, but on the sixth move, JOSS selected one of its probabilistic defections. On the next move JOSS cooperated again, but TIT FOR TAT defected in response to JOSS's previous defection. Then JOSS defected in response to TIT FOR

TABLE 1
Illustrative Game Between TIT FOR TAT and JOSS

moves	1–20	11111	23232	32323	23232
moves	21–40	32324	44444	44444	44444
moves	41–60	44444	44444	44444	44444
moves	61–80	44444	44444	44444	44444
moves	81–100	44444	44444	44444	44444
moves	101–120	44444	44444	44444	44444
moves	121–140	44444	44444	44444	44444
moves	141–160	44444	44444	44444	44444
moves	161–180	44444	44444	44444	44444
moves	181–200	44444	44444	44444	44444

Score in this game: TIT FOR TAT 236; JOSS 241.
Legend: 1 both cooperated
2 TIT FOR TAT only cooperated
3 JOSS only cooperated
4 neither cooperated

TAT's defection. In effect, the single defection of JOSS on the sixth move created an *echo* back and forth between JOSS and TIT FOR TAT. This echo resulted in JOSS defecting on all the subsequent even numbered moves and TIT FOR TAT defecting on all the subsequent odd numbered moves.

On the twenty-fifth move, JOSS selected another of its probabilistic defections. Of course, TIT FOR TAT defected on the very next move and another reverberating echo began. This echo had JOSS defecting on the odd numbered moves. Together these two echoes resulted in both players defecting on every move after move 25. This string of mutual defections meant that for the rest of the game they both got only one point per turn. The final score of this game was 236 for TIT FOR TAT and 241 for JOSS. Notice that while JOSS did a little better than TIT FOR TAT, both did poorly.[3]

The problem was a combination of an occasional defection after the other's cooperation by JOSS, combined with a short-term lack of forgiveness by both sides. The moral is that if both sides retaliate in the way that JOSS and TIT FOR TAT did, it does not pay to be as greedy as JOSS was.

A major lesson of this tournament is the importance of minimizing echo effects in an environment of mutual power. When a single defection can set off a long string of recriminations and counterrecriminations, both sides suffer. A sophisticated analysis of choice must go at least three levels deep to take account of these echo effects. The first level of analysis is the direct effect of a choice. This is easy, since a defection always earns more than a cooperation. The second level considers the indirect effects, taking into account that the other side may or may not punish a defection. This much of the analysis was certainly appreciated by many of the entrants. But the third level goes deeper and takes into account the fact that in responding to the defections of the other side, one may be repeating or even amplifying one's own previous exploitative choice. Thus a single defection may be successful when analyzed for its direct effects, and perhaps even when its secondary effects are taken into account. But the real costs may be in the tertiary effects when one's own isolated defections turn into unending mutual recriminations. Without their realizing it, many of these rules actually wound up punishing themselves. With the other player serving as a mechanism to delay the self-punishment by a few moves, this aspect of self-punishment was not picked up by many of the decision rules.

Despite the fact that none of the attempts at more or less sophisticated decision rules was an improvement on TIT FOR TAT, it was easy to find several rules that would have performed substantially better than TIT FOR TAT in the environment of the tournament. The existence of these rules should serve as a warning against the facile belief that an eye for an eye is necessarily the best strategy. There are at least three rules that would have won the tournament if submitted.

The sample program sent to prospective contestants to show them how to make a submission would in fact have won the tournament if anyone had simply clipped it and mailed it in! But no one did. The sample program defects only if the other player defected on the previous two moves. It is a more forgiving version of TIT FOR TAT in that it does not punish isolated defections. The excellent performance of this TIT FOR TWO TATS rule highlights the fact that a common error of the contestants was to expect that gains could be made from being relatively less forgiving than TIT FOR TAT, whereas in fact there were big gains to be made from being even more forgiving. The implication of this finding is striking, since it suggests that even expert strategists do not give sufficient weight to the importance of forgiveness.

Another rule which would have won the tournament was also available to most of the contestants. This was the rule which won the preliminary tournament, a report of which was used in recruiting the contestants. Called LOOK AHEAD, it was inspired by techniques used in artificial intelligence programs to play chess. It is interesting that artificial intelligence techniques could have inspired a rule which was in fact better than any of the rules designed by game theorists specifically for the Prisoner's Dilemma.

A third rule which would have won the tournament was a slight modification of DOWNING. If DOWNING had started with initial assumptions that the other players would be responsive rather than unresponsive, it too would have won and won by a large margin. A kingmaker could have been king. DOWNING's initial assumptions about the other players were pessimistic. It turned out that optimism about their responsiveness would not only have been more accurate but would also have led to more successful performance. It would have resulted in first place rather than tenth place.[4]

These results from supplementary rules reinforce a theme from the analysis of the tournament entries themselves: the entries were too competitive for their own good. In the first place, many of them defected early in the game without provocation, a characteristic which was very costly in the long run. In the second place, the optimal amount of forgiveness was considerably greater than displayed by any of the entries (except possibly DOWNING). And in the third place, the entry that was most different from the others, DOWNING, floundered on its own misplaced pessimism regarding the initial responsiveness of the others.

The analysis of the tournament results indicate that there is a lot to be learned about coping in an environment of mutual power. Even expert strategists from political science, sociology, economics, psychology, and mathematics made the systematic errors of being too competitive for their own good, not being forgiving enough, and being too pessimistic about the responsiveness of the other side.

The effectiveness of a particular strategy depends not only on its own characteristics, but also on the nature of the other strategies with which it must interact. For this reason, the results of a single tournament are not definitive. Therefore, a second round of the tournament was conducted.

The results of the second round provide substantially

better grounds for insight into the nature of effective choice in the Prisoner's Dilemma. The reason is that the entrants to the second round were all given the detailed analysis of the first round, including a discussion of the supplemental rules that would have done very well in the environment of the first round. Thus they were aware not only of the outcome of the first round, but also of the concepts used to analyze success, and the strategic pitfalls that were discovered. Moreover, they each knew that the others knew these things. Therefore, the second round presumably began at a much higher level of sophistication than the first round, and its results could be expected to be that much more valuable as a guide to effective choice in the Prisoner's Dilemma.

The second round was also a dramatic improvement over the first round in sheer size of the tournament. The response was far greater than anticipated. There was a total of sixty-two entries from six countries. The contestants were largely recruited through announcements in journals for users of small computers. The game theorists who participated in the first round of the tournament were also invited to try again. The contestants ranged from a ten-year-old computer hobbyist to professors of computer science, physics, economics, psychology, mathematics, sociology, political science, and evolutionary biology. The countries represented were the United States, Canada, Great Britain, Norway, Switzerland, and New Zealand.

The second round provided a chance both to test the validity of the themes developed in the analysis of the first round and to develop new concepts to explain successes and failures. The entrants also drew their own lessons from the experience of the first round. But different people drew different lessons. What is particularly illuminating in the second round is the way the entries based on different lessons actually interact.

TIT FOR TAT was the simplest program submitted in the first round, and it won the first round. It was the simplest submission in the second round, and it won the second round. Even though all the entrants to the second round knew that TIT FOR TAT had won the first round, no one was able to design an entry that did any better.

This decision rule was known to all of the entrants to the second round because they all had the report of the earlier round, showing that TIT FOR TAT was the most successful rule so far. They had read the arguments about how it was known to elicit a good degree of cooperation when played with humans, how it is not very exploitable, how it did well in the preliminary tournament, and how it won the first round. The report on the first round also explained some of the reasons for its success, pointing in particular to its property of never being the first to defect ("niceness") and its propensity to cooperate after the other player defected ("forgiveness" with the exception of a single punishment).

Even though an explicit tournament rule allowed anyone to submit any program, even one authored by someone else, only one person submitted TIT FOR TAT. This was Anatol Rapoport, who submitted it the first time.

The second round of the tournament was conducted in the same manner as the first round, except that minor endgame effects were eliminated. As announced in the rules, the length of the games was determined probabilistically with a 0.00346 chance of ending with each given move.[5] This is equivalent to setting $w = .99654$. Since no one knew exactly when the last move would come, end-game effects were successfully avoided in the second round.

Once again, none of the personal attributes of the contestants correlated significantly with the performance of the rules. The professors did not do significantly better than the others, nor did the Americans. Those who wrote in FORTRAN rather than BASIC did not do significantly better either, even though the use of FORTRAN would usually indicate access to something more than a bottom-of-the-line microcomputer. The names of the contestants are shown in the order of their success in appendix A along with some information about them and their programs.

On average, short programs did not do significantly better than long programs, despite the victory of TIT FOR TAT. But on the other hand, neither did long programs (with their greater complexity) do any better than short programs.

The determination of what does account for success in the second round is not easy because there were 3969 ways the 63 rules (including RANDOM) were paired in the round robin tournament. This very large tournament score matrix is given in Appendix A along with information about the entrants and their programs. In all, there were over a million moves in the second round.

As in the first round, it paid to be nice. Being the first to defect was usually quite costly. More than half of the entries were nice, so obviously most of the contestants got the message from the first round that it did not pay to be the first to defect.

In the second round, there was once again a substantial correlation between whether a rule was nice and how well it did. Of the top fifteen rules, all but one were nice (and that one ranked eighth). Of the bottom fifteen rules, all but one were not nice. The overall correlation between whether a rule was nice and its tournament score was a substantial .58.

A property that distinguishes well among the nice rules themselves is how promptly and how reliably they responded to a challenge by the other player. A rule can be called *retaliatory* if it immediately defects after an "uncalled for" defection from the other. Exactly what is meant by "uncalled for" is not precisely determined. The point, however, is that unless a strategy is incited to an immediate response by a challenge from the other player, the other player may simply take more and more frequent advantage of such an easygoing strategy.

There were a number of rules in the second round of the tournament that deliberately used controlled numbers of defections to see what they could get away with. To a large extent, what determined the actual rankings of the nice rules was how well they were able to cope with these challengers. The two challengers that were especially important in this regard I shall called TESTER and TRANQUILIZER.

TESTER was submitted by David Gladstein and came in forty-sixth in the tournament. It is designed to look for softies, but is prepared to back off if the other player shows it won't be exploited. The rule is unusual in that it defects on the very first move in order to test the other's response. If the other player ever defects, it apologizes by cooperating and playing tit-for-tat for the rest of the game. Otherwise, it cooperates on the second and third moves but defects every other move after that. TESTER did a good job of exploiting several supplementary rules that would have done quite well in the environment of the first round of the tournament. For example, TIT FOR TWO TATS defects only after the other player defects on the preceding two moves. But TESTER never does defect twice in a row. So TIT FOR TWO TATS always cooperates with TESTER, and gets badly exploited for its generosity. Notice that TESTER itself did not do particularly well in the tournament. It did, however, provide low scores for some of the more easygoing rules.

As another example of how TESTER causes problems for some rules which had done well in the first round, consider the three variants of Leslie Downing's outcome maximization principle. There were two separate submissions of the REVISED DOWNING program, based on DOWNING, which looked so promising in round one. These came from Stanley F. Quayle and Leslie Downing himself. A slightly modified version came from a youthful competitor, eleven-year-old Steve Newman. However, all three were exploited by TESTER since they all calculated that the best thing to do with a program that cooperated just over half the time after one's own cooperation was to keep on cooperating. Actually they would have been better off doing what TIT FOR TAT and many other high-ranking programs did, which was to defect immediately on the second move in response to TESTER's defection on the first move. This would have elicited TESTER's apology and things would have gone better thereafter.

TRANQUILIZER illustrates a more subtle way of taking advantage of many rules, and hence a more subtle challenge. It first seeks to establish a mutually rewarding relationship with the other player, and only then does it cautiously try to see if it will be allowed to get away with something. TRANQUILIZER was submitted by Craig

Feathers and came in twenty-seventh in the tournament. The rule normally cooperates but is ready to defect if the other player defects too often. Thus the rule tends to cooperate for the first dozen or two dozen moves if the other player is cooperating. Only then does it throw in an unprovoked defection. By waiting until a pattern of mutual cooperation has been developed, it hopes to lull the other side into being forgiving of occasional defections. If the other player continues to cooperate, then defections become more frequent. But as long as TRANQUILIZER is maintaining an average payoff of at least 2.25 points per move, it does not defect twice in succession, and it does not defect more than one-quarter of the time. It tries to avoid pressing its luck too far.

What it takes to do well with challenging rules like TESTER and TRANQUILIZER is to be ready to retaliate after an "uncalled for" defection from the other. So while it pays to be nice, it also pays to be retaliatory. TIT FOR TAT combines these desirable properties. It is nice, forgiving, and retaliatory. It is never the first to defect; it forgives an isolated defection after a single response; but it is always incited by a defection no matter how good the interaction has been so far.

The lessons of the first round of the tournament affected the environment of the second round, since the contestants were familiar with the results. The report on the first round of the Computer Prisoner's Dilemma Tournament (Axelrod 1980*a*) concluded that it paid to be not only nice but also forgiving. The contestants in the second round knew that such forgiving decision rules as TIT FOR TWO TATS and REVISED DOWNING would have done even better than TIT FOR TAT in the environment of the first round.

In the second round, many contestants apparently hoped that these conclusions would still be relevant. Of the sixty-two entries, thirty-nine were nice, and nearly all of them were at least somewhat forgiving. TIT FOR TWO TATS itself was submitted by an evolutionary biologist from the United Kingdom, John Maynard Smith. But it came in only twenty-fourth. As mentioned earlier, REVISED DOWNING was submitted twice. But in the second round, it was in the bottom half of the tournament.

What seems to have happened is an interesting interaction between people who drew one lesson and people who drew another from the first round. Lesson One was: "Be nice and forgiving." Lesson Two was more exploitative: "If others are going to be nice and forgiving, it pays to try to take advantage of them." The people who drew Lesson One suffered in the second round from those who drew Lesson Two. Rules like TRANQUILIZER and TESTER were effective at exploiting rules which were too easygoing. But the people who drew Lesson Two did not themselves do very well either. The reason is that in trying to exploit other rules, they often eventually got punished enough to make the whole game less rewarding for *both* players than pure mutual cooperation would have been. For example, TRANQUILIZER and TESTER themselves achieved only twenty-seventh and forty-sixth place, respectively. Each surpassed TIT FOR TAT's score with fewer than one-third of the rules. None of the other entries that tried to apply the exploitative conclusion of Lesson Two ranked near the top either.

While the use of Lesson Two tended to invalidate Lesson One, no entrants were able to benefit more than they were hurt in the tournament by their attempt to exploit the easygoing rules. The most successful entries tended to be relatively small variations on TIT FOR TAT which were designed to recognize and give up on a seemingly RANDOM player or a very uncooperative player. But the implementations of these ideas did not do better than the pure form of TIT FOR TAT. So TIT FOR TAT, which got along with almost everyone, won the second round of the tournament just as it had won the first round.

Would the results of the second round have been much different if the distribution of entries had been substantially different? Put another way, does TIT FOR TAT do well in a wide variety of environments? That is to say, is it *robust?*

A good way to examine this question is to construct a series of hypothetical tournaments, each with a very different distribution of the types of rules participating. The method of constructing these drastically modified tournaments is explained in appendix A. The results were that TIT FOR TAT won five of the six major variants of the tournament, and came in second in the sixth. This is a strong test of how robust the success of TIT FOR TAT really is.

Another way to examine the robustness of the results is to construct a whole sequence of hypothetical future rounds of the tournament. Some of the rules were so unsuccessful that they would be unlikely to be tried again in future tournaments, while others were successful enough that their continued presence in later tournaments would be likely. For this reason, it would be helpful to analyze what would happen over a series of tournaments if the more successful rules became a larger part of the environment for each rule, and the less successful rules were met less often. This analysis would be a strong test of a rule's performance, because continued success would require a rule to do well with other successful rules.

Evolutionary biology provides a useful way to think about this dynamic problem (Trivers 1971; Dawkins 1976, pp. 197–202; Maynard Smith 1978). Imagine that there are many animals of a single species which interact with each other quite often. Suppose the interactions take the form of a Prisoner's Dilemma. When two animals meet, they can cooperate with each other, not cooperate with each other, or one animal could exploit the other. Suppose further that each animal can recognize individuals it has already interacted with and can remember salient aspects of their interaction, such as whether the other has usually cooperated. A round of the tournament can then be regarded as a simulation of a single generation of such animals, with each decision rule being employed by large numbers of individuals. One convenient implication of this interpretation is that a given animal can interact with another animal using its own decision rule, just as it can run into an animal using some other rule.

The value of this analogy is that it allows a simulation of future generations of a tournament. The idea is that the more successful entries are more likely to be submitted in the next round, and the less successful entries are less likely to be submitted again. To make this precise, we can say that the number of copies (or offspring) of a given entry will be proportional to that entry's tournament score. We simply have to interpret the average payoff received by an individual as proportional to the individual's expected number of offspring. For example, if one rule gets twice as high a tournament score in the initial round as another rule, then it will be twice as well-represented in the next round.[6] Thus, RANDOM, for example, will be less important in the second generation, whereas TIT FOR TAT and the other high-ranking rules will be better represented.

In human terms, a rule which was not scoring well might be less likely to appear in the future for several different reasons. One possibility is that a player will try different strategies over time, and then stick with what seems to work best. Another possibility is that a person using a rule sees that other strategies are more successful and therefore switches to one of those strategies. Still another possibility is that a person occupying a key role, such as a member of Congress or the manager of a business, would be removed from that role if the strategy being followed was not very successful. Thus, learning, imitation, and selection can all operate in human affairs to produce a process which makes relatively unsuccessful strategies less likely to appear later.

The simulation of this process for the Prisoner's Dilemma tournament is actually quite straightforward. The tournament matrix gives the score each strategy gets with each of the other strategies. Starting with the proportions of each type in a given generation, it is only necessary to calculate the proportions which will exist in the next generation.[7] The better a strategy does, the more its representation will grow.

The results provide an interesting story. The first thing that happens is that the lowest-ranking eleven entries fall to half their initial size by the fifth generation while the middle-ranking entries tend to hold their own and the top-ranking entries slowly grow in size. By the fiftieth generation, the rules that ranked in the bottom third of the tournament have virtually disappeared, while most of those in the middle third have started to shrink, and those in the top third are continuing to grow (see figure 2).

This process simulates survival of the fittest. A rule that is successful on average with the current distribution of

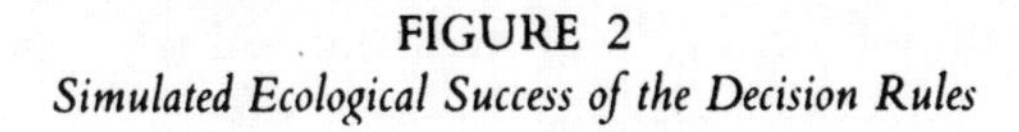

FIGURE 2
Simulated Ecological Success of the Decision Rules

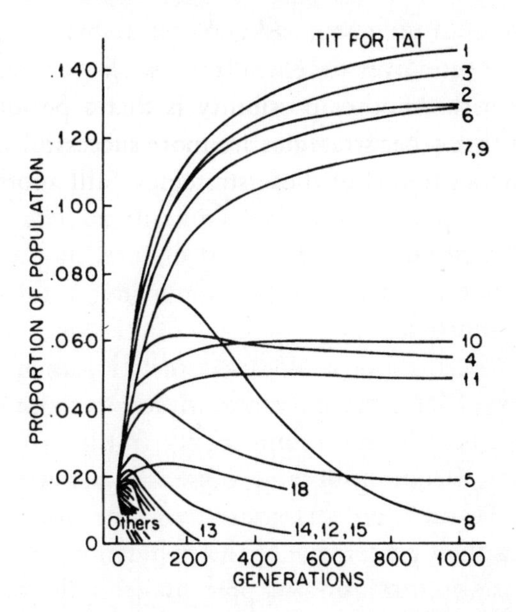

rules in the population will become an even larger proportion of the environment of the other rules in the next generation. At first, a rule that is successful with all sorts of rules will proliferate, but later as the unsuccessful rules disappear, success requires good performance with other successful rules.

This simulation provides an ecological perspective because there are no new rules of behavior introduced. It differs from an evolutionary perspective, which would allow mutations to introduce new strategies into the environment. In the ecological perspective there is a changing distribution of given types of rules. The less successful rules become less common and the more successful rules proliferate. The statistical distribution of types of individuals changes in each generation, and this changes the environment with which each of the individual types has to interact.

At first, poor programs and good programs are represented in equal proportions. But as time passes, the poorer ones begin to drop out and the good ones thrive. Success breeds more success, provided that the success derives from interactions with other successful rules. If, on the other hand, a decision rule's success derives from its ability to exploit other rules, then as these exploited rules die out, the exploiter's base of support becomes eroded and the exploiter suffers a similar fate.

A good example of ecological extinction is provided by HARRINGTON, the only non-nice rule among the top fifteen finishers in the second round. In the first two hundred or so generations of the ecological tournament, as TIT FOR TAT and the other successful nice programs were increasing their percentage of the population, HARRINGTON was also increasing its percentage. This was because of HARRINGTON's exploitative strategy. By the two hundredth generation or so, things began to take a noticeable turn. Less successful programs were becoming extinct, which meant that there were fewer and fewer prey for HARRINGTON to exploit. Soon HARRINGTON could not keep up with the successful nice rules, and by the one thousandth generation HARRINGTON was as extinct as the exploitable rules on which it preyed.

The ecological analysis shows that doing well with rules that do not score well themselves is eventually a self-defeating process. Not being nice may look promising at first, but in the long run it can destroy the very environment it needs for its own success.

The results also provide yet another victory for TIT FOR TAT. TIT FOR TAT had a very slight lead in the original tournament, and never lost this lead in simulated generations. By the one-thousandth generation it was the most successful rule and still growing at a faster rate than any other rule.

The overall record of TIT FOR TAT is very impressive. To recapitulate, in the second round, TIT FOR TAT achieved the highest average score of the sixty-two entries in the tournament. It also achieved the highest score in five of the six hypothetical tournaments which were constructed by magnifying the effects of different types of rules from the second round. And in the sixth hypothetical tournament it came in second. Finally, TIT FOR TAT never lost its first-place standing in a simulation of future generations of the tournament. Added to its victory in the first round of the tournament, and its fairly good performance in laboratory experiments with human subjects, TIT FOR TAT is clearly a very successful strategy.

Proposition 1 says that there is no absolutely best rule independent of the environment. What can be said for the empirical successes of TIT FOR TAT is that it is a very robust rule: it does very well over a wide range of environments. Part of its success might be that other rules anticipate its presence and are designed to do well with it. Doing well with TIT FOR TAT requires cooperating with it, and this in turn helps TIT FOR TAT. Even rules like TESTER that were designed to see what they could get away with, quickly apologize to TIT FOR TAT. Any rule which tries to take advantage of TIT FOR TAT will simply hurt itself. TIT FOR TAT benefits from its own nonexploitability because three conditions are satisfied:

1. The possibility of encountering TIT FOR TAT is salient.
2. Once encountered, TIT FOR TAT is easy to recognize.

3. Once recognized, TIT FOR TAT's nonexploitability is easy to appreciate.

Thus TIT FOR TAT benefits from its own *clarity.*

On the other hand, TIT FOR TAT foregoes the possibility of exploiting other rules. While such exploitation is occasionally fruitful, over a wide range of environments the problems with trying to exploit others are manifold. In the first place, if a rule defects to see what it can get away with, it risks retaliation from the rules that are provocable. In the second place, once mutual recriminations set in, it can be difficult to extract oneself. And, finally, the attempt to identify and give up on unresponsive rules (such as RANDOM or excessively uncooperative rules) often mistakenly led to giving up on rules which were in fact salvageable by a more patient rule like TIT FOR TAT. Being able to exploit the exploitable without paying too high a cost with the others is a task which was not successfully accomplished by any of the entries in round two of the tournament.

What accounts for TIT FOR TAT's robust success is its combination of being nice, retaliatory, forgiving, and clear. Its niceness prevents it from getting into unnecessary trouble. Its retaliation discourages the other side from persisting whenever defection is tried. Its forgiveness helps restore mutual cooperation. And its clarity makes it intelligible to the other player, thereby eliciting long-term cooperation.

CHAPTER 3

The Chronology of Cooperation

THE TOURNAMENT APPROACH of the previous chapter explored what happens when a given individual is likely to interact with many other players using any one of a great variety of different strategies. The results were a very clear success for TIT FOR TAT. Moreover, the ecological analysis which simulated future rounds of the tournament suggested that TIT FOR TAT would continue to thrive, and that eventually it might be used by virtually everyone.

What would happen then? Suppose that everyone came to be using the same strategy. Would there be any reason for someone to use a different strategy, or would the popular strategy remain the choice of all?

A very useful approach to this question has been developed by an evolutionary biologist, John Maynard Smith

(1974 and 1978). This approach imagines the existence of a whole population of individuals employing a certain strategy, and a single mutant individual employing a different strategy. The mutant strategy is said to invade the population if the mutant can get a higher payoff than the typical member of the population gets. Put in other terms, the whole population can be imagined to be using a single strategy, while a single individual enters the population with a new strategy. The newcomer will then be interacting only with individuals using the native strategy. Moreover, a native will almost certainly be interacting with another native since the single newcomer is a negligible part of the population. Therefore a new strategy is said to *invade* a native strategy if the newcomer gets a higher score with a native than a native gets with another native. Since the natives are virtually the entire population, the concept of invasion is equivalent to the single mutant individual being able to do better than the population average. This leads directly to the key concept of the evolutionary approach. A strategy is *collectively stable* if no strategy can invade it.[1]

The biological motivation for this approach is based on the interpretation of the payoffs in terms of fitness (survival and number of offspring). All mutations are possible; and if any could invade a given population, this mutation presumably would have the chance to do so. For this reason, only a collectively stable strategy is expected to be able to maintain itself in the long-run equilibrium as the strategy used by all. Biological applications will be discussed in chapter 5, but for now the point is that collectively stable strategies are important because they are the only ones that an entire population can maintain in the long run in the face of any possible mutant.

The motivation of applying collective stability to the

analysis of people's behavior is to discover which kinds of strategies can be maintained by a group in the face of any possible alternative strategy. If a successful alternative strategy exists, it may be found by the "mutant" individual through conscious deliberation, or through trial and error, or through just plain luck. If everyone is using a given strategy and some other strategy can do better in the environment of the current population, then someone is sure to find this better strategy sooner or later. Thus only a strategy that cannot be invaded can maintain itself as the strategy used by all.

A warning is in order about this definition of a collectively stable strategy. It assumes that the individuals who are trying out novel strategies do not interact too much with one another.[2] As will be shown further on, if they do interact in clusters, then new and very important developments are possible.

A difficulty with this concept of collective stability when applied to the iterated Prisoner's Dilemma is that it can be very hard actually to determine which strategies have it and which do not. Others have dealt with this difficulty by restricting the analysis to situations where the strategies are particularly simple, or by considering only some arbitrarily limited set of strategies.[3] The problem has now been solved, making it possible to characterize *all* collectively stable strategies for the iterated Prisoner's Dilemma. The characterization is given in Appendix B.

For present purposes, it is not necessary to be so general. It is sufficient to take a particular strategy and see under what conditions it can resist invasion by any other strategy. A good strategy to investigate is TIT FOR TAT. TIT FOR TAT cooperates on the first move, and then does whatever the other player did on the previous move. Thus a population of players using TIT FOR TAT will cooperate with each other, and each will get *R* per move. If another strategy is to invade this population, it must get a higher expected value than this. What kind of strategy might be able to get more than this when playing with a player using TIT FOR TAT?

The first thing that can be said is that such a strategy must defect at some point, since otherwise it will get *R* per move just as the others do. When it first defects it will get the temptation, *T*, which is the highest payoff. But then TIT FOR TAT will defect. Consequently, TIT FOR TAT can avoid being invaded by such a rule only if the game is likely to last long enough for the retaliation to counteract the temptation to defect. In fact, no rule can invade TIT FOR TAT if the discount parameter, *w*, is sufficiently large.

The way to demonstrate this is to use the fact that TIT FOR TAT has a memory of only one move. Therefore, an effective challenger can take maximum advantage of it by repeating whatever sequence of choices of cooperation and defection work best. Because of the short memory, the sequence to be repeated need be no longer than two moves. Thus the most effective challengers will be repeated sequences of DC, or DD (which is ALL D). If neither of these strategies can invade TIT FOR TAT, then no strategy can, and TIT FOR TAT is collectively stable.

Both of these potential challengers do better than *R* on the first move, and do worse than *R* on the second move. So they benefit from settings where the future is not too important relative to the present. However, if *w* is large enough, neither ALL D nor the alternation of D and C can invade TIT FOR TAT. And if neither of these two strategies can invade TIT FOR TAT, then no other strategies can. This gives the second proposition. The proof is in Appendix B.

Proposition 2. TIT FOR TAT is collectively stable if and only if, w is large enough. This critical value of w is a function of the four payoff parameters, T, R, P, and S.[4]

The significance of this proposition is that if everyone in a population is cooperating with everyone else because each is using the TIT FOR TAT strategy, no one can do better using any other strategy *providing* that the future casts a large enough shadow onto the present. In other words, what makes it impossible for TIT FOR TAT to be invaded is that the discount parameter, w, is high enough relative to the requirement determined by the four payoff parameters. For example, suppose that $T=5$, $R=3$, $P=1$, and $S=0$ as in the payoff matrix shown in figure 1. Then TIT FOR TAT is collectively stable if the next move is at least ⅔ as important as the current move. Under these conditions, if everyone else is using TIT FOR TAT, you can do no better than to do the same, and cooperate with them. On the other hand, if w falls below this critical value, and everyone else is using TIT FOR TAT, it will pay to defect on alternative moves. If w is less than ½, it even pays to always defect.

One specific implication is that if the other player is unlikely to be around much longer because of apparent weakness, then the perceived value of w falls and the reciprocity of TIT FOR TAT is no longer stable. We have Caesar's explanation of why Pompey's allies stopped cooperating with him. "They regarded his [Pompey's] prospects as hopeless and acted according to the common rule by which a man's friends become his enemies in adversity" [translated by Warner 1960, p. 328].

Another example is the case where a business is on the

edge of bankruptcy and sells its accounts receivable to an outsider called a factor. This sale is made at a very substantial discount because

> once a manufacturer begins to go under, even his best customers begin refusing payment for merchandise, claiming defects in quality, failure to meet specifications, tardy delivery, or what-have-you. The great enforcer of morality in commerce is the continuing relationship, the belief that one will have to do business again with this customer, or this supplier, and when a failing company loses this automatic enforcer, not even a strong-arm factor is likely to find a substitute. (Mayer, 1974, p. 280)

Similarly, any member of Congress who is perceived as likely to be defeated in the next election may have some difficulty doing legislative business with colleagues on the usual basis of trust and good credit.[5]

There are many other examples of the importance of long-term interaction for the stability of cooperation. It is easier to maintain the norms of reciprocity in a stable small town or ethnic neighborhood. Conversely, a visiting professor is likely to receive poor treatment by other faculty members compared to the way these same people treat their regular colleagues.

A fascinating case of the development of cooperation based on continuing interaction occurred in the trench warfare of World War I. In the midst of this very brutal war there developed between the men facing each other what came to be called the "live-and-let-live system." The troops would attack each other when ordered to do so, but between large battles each side would deliberately avoid doing much harm to the other side—provided that the other side reciprocated. The strategy was not necessarily TIT FOR TAT. Sometimes it was two for one. As a British officer wrote in his memoirs of the takeover of a new sector from the French:

> It was the French practice to "let sleeping dogs lie" when in a quiet sector . . . and of making this clear by retorting vigorously only when challenged. In one sector which we took over from them they explained to me that they had practically a code which the enemy well understood: they fired two shots for every one that came over, but never fired first. (Kelly 1930, p. 18)

Such practices of tacit cooperation were quite illegal—but they were also endemic. For several years this system developed and elaborated itself despite the passions of the war and the best efforts of the generals to pursue a policy of constant attrition. The story is so rich in illuminating detail that all of the next chapter will be devoted to it.

Even without going further into the episode of trench warfare, the occurrence of a two-for-one strategy does suggest that one must be careful about drawing conclusions from a narrow focus on a pure TIT FOR TAT strategy. Just how broadly applicable was the proposition about TIT FOR TAT which said that it was collectively stable if and only if the future of the interaction was sufficiently important? The next proposition says that this result is very general indeed, and actually applies to any strategy which may be the first to cooperate.

Proposition 3. Any strategy which may be the first to cooperate can be collectively stable only when w is sufficiently large.

The reason is that for a strategy to be collectively stable it must protect itself from invasion by any challenger, including the strategy which always defects. If the native strategy ever cooperates, ALL D will get T on that move. On the other hand, the population average among the natives can be no greater than *R* per move. So in order for the population average to be no less than the score of the challenging ALL D, the interaction must last long enough for the gain from temptation to be nullified over future moves. This is the heart of the matter, but for the formal proof see appendix B.

The TIT FOR TAT and the two-for-one strategies are both *nice* decision rules in that they are never the first to defect. The advantage of a nice rule in resisting invasion is that it attains the highest score possible in a population consisting of a single type of strategy. It does this by getting the reward for mutual cooperation on each move with another player using the same strategy.

The TIT FOR TAT and the two-for-one strategy share something else as well. They both retaliate after a defection by the other. This observation leads to a general principle, since any collectively stable strategy which is willing to cooperate must somehow make it unprofitable for a challenger to try to exploit it. The general principle is that a nice rule must be *provoked* by the very first defection of the other player, meaning that on some later move the strategy must have a finite chance of responding with a defection of its own.[6]

Proposition 4. For a nice strategy to be collectively stable, it must be provoked by the very first defection of the other player.

The reason is simple enough. If a nice strategy were not provoked by a defection on move *n,* then it would not be collectively stable because it could be invaded by a rule which defected only on move *n.*

These last two propositions have shown that a nice rule can be collectively stable if the future casts a large enough shadow and the rule itself is provocable. But there is one strategy which is *always* collectively stable, regardless of the value of the discount parameter, w, or the payoff parameters, T, R, P, and S. This is ALL D, the rule which defects no matter what.

Proposition 5. ALL D is always collectively stable.

If the other player is certain to defect, there is no point in your ever cooperating. A population of players using ALL D will each get P per move. There is no way a player can do any better than this if no one else will ever cooperate. After all, any cooperative choice would just yield the sucker's payoff, S, with no chance for future compensation.

This proposition has important implications for the evolution of cooperation. If one imagines a system starting with individuals who cannot be enticed to cooperate, the collective stability of ALL D implies that no single individual can hope to do any better than go along and be uncooperative as well. A world of "meanies" can resist invasion by anyone using any other strategy—provided that the newcomers arrive one at a time. The problem, of course, is that a single newcomer in such a mean world has no one who will reciprocate any cooperation. If the newcomers arrive in small clusters, however, they will have a chance to get cooperation started.

To see how this can happen, consider a simple numerical example with the payoff matrix given in figure 1 on page 8. This example sets the temptation to exploit at $T=5$, the reward for mutual cooperation at $R=3$, the punishment for mutual defection at $P=1$, and the sucker's payoff at $S=0$. Further, suppose that the probability of two players meeting again is $w=.9$. Then in a population of meanies using ALL D, each will each get a payoff of P each turn for a cumulative score of 10 points.

Now suppose several players use TIT FOR TAT. When

a TIT FOR TAT interacts with an ALL D, the TIT FOR TAT will be exploited on the first move, and then will not cooperate again with a meanie. This gives 0 on the first move and 1 on subsequent moves, for a cumulative score of 9 points.[7] This score is a little less than the 10 points that the meanies get with each other. However, if TIT FOR TAT interacts with another TIT FOR TAT, they achieve mutual cooperation from the start and both get 3 points each move which cumulates to 30 points. This score is much better than the 10 points that the meanies get with each other.

Now if the TIT FOR TAT newcomers are a negligible proportion of the entire population, the meanies will be almost always interacting with other meanies and getting only 10 points. So if the TIT FOR TAT players can interact with each other enough, they can achieve a higher average score than this 10 points. They can do so by having enough opportunities to score 30 points with someone who will reciprocate their cooperation rather than 9 points with someone who won't. How much will it take? If a TIT FOR TAT has some proportion, p, of its interactions with other TIT FOR TAT players, it will have $1-p$ with the meanies. So its average score will be $30p + 9(1-p)$. If this score is more than 10 points, it pays to use a TIT FOR TAT strategy rather than be a meanie like the bulk of the population. And this will be true even if only 5 percent of the interactions of the TIT FOR TAT players are with other TIT FOR TAT players.[8] Thus, even a small cluster of TIT FOR TAT players can get a higher average score than the large population of meanies they enter. Because the TIT FOR TAT players do so well when they do meet each other, they do not have to meet each other very often to make their strategy the superior one to use.

In this way, a world of meanies can be invaded by a cluster of TIT FOR TAT—and rather easily at that. To illustrate this point, suppose a business school teacher taught a class of students to initiate cooperative behavior in the firms they join, and to reciprocate cooperation from other firms. If the students did act this way, and if they did not disperse too widely (so that a sufficient proportion of their interactions were with other members of the same class), then the students would find that their lessons paid off. In the numerical example just discussed, a firm switching to TIT FOR TAT would need to have only 5 percent of its interactions with another such firm for them to be glad they gave cooperation a chance.

Even less clustering is necessary when the interactions are expected to be of longer duration or the time discount factor is not as great. Using the interpretation of w as reflecting the chance of meeting once again, suppose the median game length is two hundred moves (corresponding to $w = .99654$). In this case even one interaction out of a thousand with a like-minded follower of TIT FOR TAT is enough for the strategy to invade a world of ALL D's. Even with a median game length of only two moves ($w = .5$), anything over a fifth of the interactions by the TIT FOR TAT players with like-minded types is sufficient for invasion to succeed and cooperation to emerge.

This concept of invasion by a cluster can be precisely defined and applied to any strategy. Suppose that a native strategy is being used by virtually everyone, and that a small group of individuals using a new strategy arrives and interacts both with the other newcomers and with the natives. The proportion of interactions by someone using the new strategy with another individual using the new strategy is p. Assuming that the newcomers are rare relative to the natives, virtually all of the interactions of natives are with other natives. Then the average score of a newcomer is the weighted average of what the newcomer gets with another newcomer, and what the newcomer gets with a native. The weights are the frequency of these two events, namely p and $1-p$. On the other hand, the average score of a native is virtually identical with what a native gets with another native since the newcomers are so rare. This reasoning allows one to establish that the cluster of newcomers can invade the natives, if the newcomers do well with other newcomers and if the newcomers meet each other sufficiently often.[9]

Notice that this assumes that pairing in the interactions is not random. With random pairing, a newcomer would rarely meet another newcomer. Instead, the clustering concept treats the case in which the newcomers are a trivial part of the environment of the natives, but a nontrivial part of the environment of the newcomers themselves.

The next result shows which strategies are the most efficient at invading ALL D with the least amount of clustering. These are the strategies which are best able to discriminate between themselves and ALL D. A strategy is *maximally discriminating* if it will eventually cooperate even if the other has never cooperated yet, and once it cooperates will never cooperate again with ALL D but will always cooperate with another player using the same strategy as it uses.

Proposition 6. The strategies which can invade ALL D in a cluster with the smallest value of p are those which are maximally discriminating, such as TIT FOR TAT.

It is easy to see that TIT FOR TAT is a maximally discriminating strategy. It cooperates on the very first move, but once it cooperates with ALL D it will never cooperate again. On the other hand, it will have an unbroken string of cooperation with another TIT FOR TAT player. So TIT FOR TAT is very good at discriminating between its own twin and an ALL D, and this property allows it to invade a world of meanies with the smallest possible cluster.

While clustering suggests a mechanism for the initiation of cooperation in a world of meanies, it also raises the question of whether the reverse could happen once a strategy like TIT FOR TAT becomes established itself. Actually, there is a surprising and very pleasant asymmetry here. To see what it is, recall the definition of a *nice* strategy as one, such as TIT FOR TAT, which will never be the first to defect. Obviously when two nice strategies interact, they both receive R each move, which is the highest average score an individual can get when interacting with another individual using the same strategy. This gives the following proposition.

Proposition 7. If a nice strategy cannot be invaded by a single individual, it cannot be invaded by any cluster of individuals either.

The score achieved by a strategy that comes in a cluster is a weighted average of two components: how it does with others of its kind and how it does with the predominant strategy. Both of these components are less than or equal to the score achieved by the predominant, nice strategy. Therefore, if the predominant, nice strategy cannot be invaded by a single individual it cannot be invaded by a cluster either.

This conclusion means nice rules do not have the structural weakness displayed in ALL D. ALL D can withstand invasion by any strategy as long as the players using these other strategies come one at a time. But if they come in clusters (even in rather small clusters), ALL D can be invaded. With nice rules, the situation is different. If a nice rule can resist invasion by other nice rules coming one at a time, then it can resist invasion by clusters, no matter how large. So nice rules can protect themselves in a way that meanies cannot.

These results fit together to give a chronological picture of the evolution of cooperation. In the illustrative case of the Senate, proposition 5 has shown that without clustering (or some comparable mechanism), the original pattern of mutual "treachery" could not have been overcome. Perhaps these critical early clusters were based on the small groups of representatives who lived together in boardinghouses in the new capital during the Jeffersonian era (Young 1966). Or perhaps the state delegations and state party delegations were more critical (Bogue and Marlaire 1975). Proposition 7 demonstrates that once cooperation based on reciprocity has become established, it can remain stable even if a cluster of newcomers does not respect this senatorial folkway. And now that the pattern of reciprocity is established, propositions 2 and 3 show that it is collectively stable, as long as the biennial turnover rate is not too great.

Thus cooperation can emerge even in a world of unconditional defection. The development cannot take place if it is tried only by scattered individuals who have no chance to interact with each other. But cooperation can emerge from small clusters of discriminating individuals, as long as these individuals have even a small proportion of their interactions with each other. Moreover, if nice strategies (those which are never the first to defect) come to be adopted by virtually everyone, then those individuals can afford to be generous in dealing with any others. By doing so well with each other, a population of nice rules can protect themselves against clusters of individuals using any other strategy just as well as they can protect themselves against single individuals. But for a nice strategy to be stable in the collective sense, it must be provocable. So mutual cooperation can emerge in a world of egoists without central control by starting with a cluster of individuals who rely on reciprocity.

To see how widely these results apply, the next two chapters explore cases in which cooperation has actually evolved. The first case is one in which cooperation evolved in wartime despite bitter antagonism between the players. The second involves biological systems in which lower animals cannot appreciate the consequences of their choices. These cases demonstrate that when the conditions are right, cooperation can evolve even without friendship or foresight.

III

Cooperation Without Friendship or Foresight

CHAPTER 4

The Live-and-Let-Live System in Trench Warfare in World War I

SOMETIMES cooperation emerges where it is least expected. During World War I, the Western Front was the scene of horrible battles for a few yards of territory. But between these battles, and even during them at other places along the five-hundred-mile line in France and Belgium, the enemy soldiers often exercised considerable restraint. A British staff officer on a tour of the trenches remarked that he was

> astonished to observe German soldiers walking about within rifle range behind their own line. Our men appeared to take no notice. I privately made up my mind to do away with that sort of thing when we took over; such things should not be allowed.

These people evidently did not know there was a war on. Both sides apparently believed in the policy of "live and let live." (Dugdale 1932, p. 94)

This is not an isolated example. The live-and-let-live system was endemic in trench warfare. It flourished despite the best efforts of senior officers to stop it, despite the passions aroused by combat, despite the military logic of kill or be killed, and despite the ease with which the high command was able to repress any local efforts to arrange a direct truce.

This is a case of cooperation emerging despite great antagonism between the players. As such, it provides a challenge for the application of the concepts and the theory developed in the first three chapters. In particular, the main goal is to use the theory to explain:

1. How could the live-and-let-live system have gotten started?
2. How was it sustained?
3. Why did it break down toward the end of the war?
4. Why was it characteristic of trench warfare in World War I, but of few other wars?

A second goal is to use the historical case to suggest how the original concepts and theory can be further elaborated.

Fortunately, a recent book-length study of the live-and-let-live system is available. This excellent work by a British sociologist, Tony Ashworth (1980), is based upon diaries, letters, and reminiscences of trench fighters. Material was found from virtually every one of the fifty-seven British divisions, with an average of more than three sources per division. To a lesser extent, material from French and German sources was also consulted. The result is a very rich set of illustrations that are analyzed with great skill to provide a comprehensive picture of the development and character of trench warfare on the Western Front in World War I. This chapter relies upon Ashworth's fine work for its illustrative quotes and for its historical interpretation.

While Ashworth does not put it this way, the historical situation in the quiet sectors along the Western Front was an iterated Prisoner's Dilemma. In a given locality, the two players can be taken to be the small units facing each other. At any time, the choices are to shoot to kill or deliberately to shoot to avoid causing damage. For both sides, weakening the enemy is an important value because it will promote survival if a major battle is ordered in the sector. Therefore, in the short run it is better to do damage now whether the enemy is shooting back or not. This establishes that mutual defection is preferred to unilateral restraint ($P>S$), and that unilateral restraint by the other side is even better than mutual cooperation ($T>R$). In addition, the reward for mutual restraint is preferred by the local units to the outcome of mutual punishment ($R>P$), since mutual punishment would imply that both units would suffer for little or no relative gain. Taken together, this establishes the essential set of inequalities: $T>R>P>S$. Moreover, both sides would prefer mutual restraint to the random alternation of serious hostilities, making $R>(T+S)/2$. Thus the situation meets the conditions for a Prisoner's Dilemma between small units facing each other in a given immobile sector.

Two small units facing each other across one hundred to four hundred yards of no-man's-land were the players in one of these potentially deadly Prisoner's Dilemmas. Typically, the basic unit could be taken to be the battalion, consisting of about one thousand men, half of whom would be in the front line at any one time. The battalion played a large role in the life of an infantryman. It not only organized its members for combat, but also fed, paid, and clothed them as well as arranged their leave. All of the officers and most of the other soldiers in the battalion knew each other by sight. For our purposes, two key factors make the battalion the most typical player. On the one hand, it was large enough to occupy a sufficient sector of the front to be "held accountable" for aggressive actions which came from its territory. On the other hand, it was small enough to be able to control the individual behavior of its men, through a variety of means, both formal and informal.

A battalion on one side might be facing parts of one, two, or three battalions on the other side. Thus each player could simultaneously be involved in several interactions. Over the course of the Western Front, there would be hundreds of such face-offs.

Only the small units were involved in these Prisoner's Dilemmas. The high commands of the two sides did not share the view of the common soldier who said:

> The real reason for the quietness of some sections of the line was that neither side had any intention of advancing in that particular district. . . . If the British shelled the Germans, the Germans replied, and the damage was equal: if the Germans bombed an advanced piece of trench and killed five Englishmen, an answering fusillade killed five Germans. (Belton Cobb 1916, p. 74)

To the army headquarters, the important thing was to develop an offensive spirit in the troops. The Allies, in particular, pursued a strategy of attrition whereby equal losses in men from both sides meant a net gain for the Allies because sooner or later Germany's strength would be exhausted first. So at the national level, World War I approximated a zero-sum game in which losses for one side represented gains for the other side. But at the local level, along the front line, mutual restraint was much preferred to mutual punishment.

Locally, the dilemma persisted: at any given moment it was prudent to shoot to kill, whether the other side did so or not. What made trench warfare so different from most other combat was that the same small units faced each other in immobile sectors for extended periods of time. This changed the game from a one-move Prisoner's Dilemma in which defection is the dominant choice, to an iterated Prisoner's Dilemma in which conditional strategies are possible. The result accorded with the theory's predictions: with sustained interaction, the stable outcome could be mutual cooperation based upon reciprocity. In particular, both sides followed strategies that would not be the first to defect, but that would be provoked if the other defected.

Before looking further at the stability of the cooperation, it is interesting to see how cooperation got started in the first place. The first stage of the war, which began in August 1914, was highly mobile and very bloody. But as the lines stabilized, nonaggression between the troops emerged spontaneously in many places along the front. The earliest instances may have been associated with meals which were served at the same times on both sides of no-man's land. As early as November 1914, a noncommissioned officer whose unit had been in the trenches for some days, observed that

> the quartermaster used to bring the rations up . . . each night after dark; they were laid out and parties used to come from the front line to fetch them. I suppose the enemy were occupied in the same way; so things were quiet at that hour for a couple of nights, and the ration parties became careless because of it, and

laughed and talked on their way back to their companies. *(The War the Infantry Knew* 1938, p. 92)

By Christmas there was extensive fraternization, a practice which the headquarters frowned upon. In the following months, direct truces were occasionally arranged by shouts or by signals. An eyewitness noted that:

In one section the hour of 8 to 9 A.M. was regarded as consecrated to "private business," and certain places indicated by a flag were regarded as out of bounds by the snipers on both sides. (Morgan 1916, pp. 270–71)

But direct truces were easily suppressed. Orders were issued making clear that the soldiers "were in France to fight and not to fraternize with the enemy" (*Fifth Battalion the Camaronians* 1936, p. 28). More to the point, several soldiers were courtmartialed and whole battalions were punished. Soon it became clear that verbal arrangements were easily suppressed by the high command and such arrangements became rare.

Another way in which mutual restraint got started was during a spell of miserable weather. When the rains were bad enough, it was almost impossible to undertake major aggressive action. Often ad hoc weather truces emerged in which the troops simply did not shoot at each other. When the weather improved, the pattern of mutual restraint sometimes simply continued.

So verbal agreements were effective in getting cooperation started on many occasions early in the war, but direct fraternization was easily suppressed. More effective in the long run were various methods which allowed the two sides to coordinate their actions without having to resort to words. A key factor was the realization that if one side would exercise a particular kind of restraint, then the other might reciprocate. Similarities in basic needs and activities let the soldiers appreciate that the other side would probably not be following a strategy of unconditional defection. For example, in the summer of 1915, a soldier saw that the enemy would be likely to reciprocate cooperation based on the desire for fresh rations.

> It would be child's play to shell the road behind the enemy's trenches, crowded as it must be with ration wagons and water carts, into a bloodstained wilderness . . . but on the whole there is silence. After all, if you prevent your enemy from drawing his rations, his remedy is simple: he will prevent you from drawing yours. (Hay 1916, pp. 224–25)

Once started, strategies based on reciprocity could spread in a variety of ways. A restraint undertaken in certain hours could be extended to longer hours. A particular kind of restraint could lead to attempting other kinds of restraint. And most importantly of all, the progress achieved in one small sector of the front could be imitated by the units in neighboring sectors.

Just as important as getting cooperation started were the conditions that allowed it to be sustainable. The strategies that could sustain mutual cooperation were the ones which were provocable. During the periods of mutual restraint, the enemy soldiers took pains to show each other that they could indeed retaliate if necessary. For example, German snipers showed their prowess to the British by aiming at spots on the walls of cottages and firing until they had cut a hole (*The War the Infantry Knew* 1938, p. 98). Likewise the artillery would often demonstrate with a few accurately aimed shots that they could do more damage if they wished. These demonstrations of retaliatory capabilities helped police the system by showing that restraint was not due to weakness, and that defection would be self-defeating.

When a defection actually occurred, the retaliation was often more than would be called for by TIT FOR TAT. Two-for-one or three-for-one was a common response to an act that went beyond what was considered acceptable.

> We go out at night in front of the trenches. . . . The German working parties are also out, so it is not considered etiquette to fire. The really nasty things are rifle grenades. . . . They can kill as many as eight or nine men if they do fall into a trench. . . . But we never use ours unless the Germans get particularly noisy, as on their system of retaliation three for every one of ours come back. (Greenwell 1972, pp. 16–17)

There was probably an inherent damping process that usually prevented these retaliations from leading to an uncontrolled echo of mutual recriminations. The side that instigated the action might note the escalated response and not try to redouble or retriple it. Once the escalation was not driven further, it would probably tend to die out. Since not every bullet, grenade, or shell fired in earnest would hit its target, there would be an inherent tendency toward de-escalation.

Another problem that had to be overcome to maintain the stability of cooperation was the rotation of troops. About every eight days, a battalion would change places with another battalion billeted behind it. At longer intervals, larger units would change places. What allowed the cooperation to remain stable was the process of familiarization that the outgoing unit would provide for the incoming unit. The particular details of the tacit understandings with the enemy were explained. But sometimes it was quite sufficient for an old timer to point out to a newcomer that "Mr. Bosche ain't a bad fellow. You leave 'im alone; 'e'll leave you alone" (Gillon n.d., p. 77). This socialization allowed one unit to pick up the game right where the other left it.

Still another problem for the maintenance of stable cooperation was the fact that the artillery was much less vulnerable to enemy retaliation than was the infantry. Therefore, the artillery had a lesser stake in the live-and-let-live system. As a consequence, the infantry tended to be solicitous of the forward observers from the artillery. As a German artillery man noted of the infantry, "If they ever have any delicacies to spare, they make us a present of them, partly of course because they feel we are protecting them" (Sulzbach 1973, p. 71). The goal was to encourage the artillery to respect the infantry's desire to let sleeping dogs lie. A new forward observer for the artillery was often greeted by the infantry with the request, "I hope you are not going to start trouble." The best answer was, "Not unless *you* want" (Ashworth 1980, p. 169). This reflected the dual role of artillery in the maintenance of mutual restraint with the enemy: the passiveness when unprovoked, and the instant retaliation when the enemy broke the peace.

The high commands of the British, French, and German armies all wanted to put a stop to tacit truces; all were afraid that they sapped the morale of their men, and all believed throughout the war that a ceaseless policy of offense was the only way to victory. With few exceptions, the headquarters could enforce any orders that they could directly monitor. Thus the headquarters were able to conduct large battles by ordering the men to leave their trenches and risk their lives in charging the enemy positions. But between large battles, they were not able to monitor their orders to keep up the pressure.[1] After all, it was hard for a senior officer to determine who was shooting to kill, and who was shooting with an eye to avoiding retaliation. The soldiers became expert at defeating the monitoring system, as when a unit kept a coil of enemy wire and sent a piece to headquarters whenever asked to prove that they had conducted a patrol of no-man's-land.

What finally destroyed the live-and-let-live system was the institution of a type of incessant aggression that the headquarters *could* monitor. This was the raid, a carefully prepared attack on enemy trenches which involved from ten to two hundred men. Raiders were ordered to kill or capture the enemy in his own trenches. If the raid was successful, prisoners would be taken; and if the raid was a failure, casualties would be proof of the attempt. There was no effective way to pretend that a raid had been undertaken when it had not. And there was no effective way to cooperate with the enemy in a raid because neither live soldiers nor dead bodies could be exchanged.

The live-and-let-live system could not cope with the disruption caused by the hundreds of small raids. After a raid neither side knew what to expect next. The side that had raided could expect retaliation but could not predict when, where, or how. The side that had been raided was also nervous, not knowing whether the raid was an isolated attack or the first of a series. Moreover, since raids could be ordered and monitored from headquarters, the magnitude of the retaliatory raid could also be controlled, preventing a dampening of the process. The battalions were forced to mount real attacks on the enemy, the retaliation was undampened, and the process echoed out of control.

Ironically, when the British High Command undertook

its policy of raiding, it did not do so in order to end the live-and-let-live system. Instead, its initial goal was political, namely, to show their French allies that they were doing their part to harass the enemy. Their image of the direct effects of raiding was that it increased the morale of their own troops by restoring an offensive spirit and that it promoted attrition by inflicting more casualties on the enemy in the raids than the raiding troops themselves would suffer. Whether these effects on morale and casualty ratios were realized has been debated ever since. What is clear in retrospect is that the indirect effect of the raids was to destroy the conditions needed for the stability of the tacit restraints widely exercised on the Western Front. Without realizing exactly what they were doing, the high command effectively ended the live-and-let-live system by preventing their battalions from exercising their own strategies of cooperation based on reciprocity.

The introduction of raids completed the cycle of the evolution of the live-and-let-live system. Cooperation got a foothold through exploratory actions at the local level, was able to sustain itself because of the duration of contact between small units facing each other, and was eventually undermined when these small units lost their freedom of action. Small units, such as battalions, used their own strategies in dealing with the enemy units they faced. Cooperation first emerged spontaneously in a variety of contexts, such as restraint in attacking the distribution of enemy rations, a pause during the first Christmas in the trenches, and a slow resumption of fighting after bad weather made sustained combat almost impossible. These restraints quickly evolved into clear patterns of mutually understood behavior, such as two-for-one or three-for-one retaliation for actions that were taken to be unacceptable. The mechanisms of the evolution of these strategies must have been trial and error and the imitation of neighboring units.

The mechanisms for evolution involved neither blind mutation nor survival of the fittest. Unlike blind mutation, the soldiers understood their situation and actively tried to make the most of it. They understood the indirect consequences of their acts as embodied in what I call the echo principle: "To provide discomfort for the other is but a roundabout way of providing it for themselves" (Sorley 1919, p. 283). The strategies were based on thought as well as experience. The soldiers learned that to maintain mutual restraint with their enemies, they had to base that restraint on a demonstrated capability and willingness to be provoked. They learned that cooperation had to be based upon reciprocity. Thus, the evolution of strategies was based on deliberate rather than blind adaptation. Nor did the evolution involve survival of the fittest. While an ineffective strategy would mean more casualties for the unit, replacements typically meant that the units themselves would survive.

The origins, maintenance, and destruction of the live-and-let-live system of trench warfare are all consistent with the theory of the evolution of cooperation. In addition, there are two very interesting developments within the live-and-let-live system which are new to the theory. These additional developments are the emergence of ethics and ritual.

The ethics that developed are illustrated in this incident, related by a British officer recalling his experience while facing a Saxon unit of the German Army.

I was having tea with A Company when we heard a lot of shouting and went out to investigate. We found our men and the

Germans standing on their respective parapets. Suddenly a salvo arrived but did no damage. Naturally both sides got down and our men started swearing at the Germans, when all at once a brave German got on to his parapet and shouted out "We are very sorry about that; we hope no one was hurt. It is not our fault, it is that damned Prussian artillery." (Rutter 1934, p. 29)

This Saxon apology goes well beyond a merely instrumental effort to prevent retaliation. It reflects moral regret for having violated a situation of trust, and it shows concern that someone might have been hurt.

The cooperative exchanges of mutual restraint actually changed the nature of the interaction. They tended to make the two sides care about each other's welfare. This change can be interpreted in terms of the Prisoner's Dilemma by saying that the very experience of sustained mutual cooperation altered the payoffs of the players, making mutual cooperation even more valued than it was before.

The converse was also true. When the pattern of mutual cooperation deteriorated due to mandatory raiding, a powerful ethic of revenge was evoked. This ethic was not just a question of calmly following a strategy based on reciprocity. It was also a question of doing what seemed moral and proper to fulfill one's obligation to a fallen comrade. And revenge evoked revenge. Thus both cooperation and defection were self-reinforcing. The self-reinforcement of these mutual behavioral patterns was not only in terms of the interacting strategies of the players, but also in terms of their perceptions of the meaning of the outcomes. In abstract terms, the point is that not only did preferences affect behavior and outcomes, but behavior and outcomes also affected preferences.

The other addition to the theory suggested by the trench warfare case is the development of ritual. The rituals took the form of perfunctory use of small arms, and deliberately harmless use of artillery. For example, the Germans in one place conducted "their offensive operations with a tactful blend of constant firing and bad shooting, which while it satisfies the Prussians causes no serious inconvenience to Thomas Atkins" (Hay 1916, p. 206).

Even more striking was the predictable use of artillery which occurred in many sectors.

So regular were they [the Germans] in their choice of targets, times of shooting, and number of rounds fired, that, after being in the line one or two days, Colonel Jones had discovered their system, and knew to a minute where the next shell would fall. His calculations were very accurate, and he was able to take what seemed to uninitiated Staff Officers big risks, knowing that the shelling would stop before he reached the place being shelled. (Hills 1919, p. 96)

The other side did the same thing, as noted by a German soldier commenting on "the evening gun" fired by the British.

At seven it came—so regularly that you could set your watch by it. . . . It always had the same objective, its range was accurate, it never varied laterally or went beyond or fell short of the mark. . . . There were even some inquisitive fellows who crawled out . . . a little before seven, in order to see it burst. (Koppen 1931, pp. 135–37)

These rituals of perfunctory and routine firing sent a double message. To the high command they conveyed aggression, but to the enemy they conveyed peace. The men pretended to be implementing an aggressive policy, but were not. Ashworth himself explains that these stylized acts were more than a way of avoiding retaliation.

In trench war, a structure of ritualised aggression was a ceremony where antagonists participated in regular, reciprocal discharges of missiles, that is, bombs, bullets and so forth, which symbolized and strengthened, at one and the same time, both sentiments of fellow-feelings, and beliefs that the enemy was a fellow sufferer. (Ashworth 1980, p. 144)

Thus these rituals helped strengthen the moral sanctions which reinforced the evolutionary basis of the live-and-let-live system.

The live-and-let-live system that emerged in the bitter trench warfare of World War I demonstrates that friendship is hardly necessary for cooperation based upon reciprocity to get started. Under suitable circumstances, cooperation can develop even between antagonists.

One thing the soldiers in the trenches had going for them was a fairly clear understanding of the role of reciprocity in the maintenance of the cooperation. The next chapter uses biological examples to demonstrate that such understanding by the participants is not really necessary for cooperation to emerge and prove stable.

CHAPTER 5

The Evolution of Cooperation in Biological Systems

(with William D. Hamilton)

IN EARLIER CHAPTERS, several concepts from evolutionary biology were borrowed to help analyze the emergence of cooperation between people. In this chapter, the favor is returned. The findings and theory that have been developed to understand people will now be applied to the analysis of cooperation in biological evolution. An important conclusion drawn from this investigation is that foresight is not necessary for the evolution of cooperation.

The theory of biological evolution is based on the struggle for life and the survival of the fittest. Yet cooperation is common between members of the same species and even between members of different species. Before about 1960, accounts of the evolutionary process largely dismissed cooperative phenomena as not requiring special attention. This dismissal followed from a misreading of theory that assigned most adaptation to selection at the level of populations or whole species. As a result of such misreading, cooperation was always considered adaptive. Recent reviews of the evolutionary process, however, have shown no sound basis for viewing selection as being based upon benefits to whole groups. Quite the contrary. At the level of a species or a population, the processes of selection are weak. The original individualistic emphasis of Darwin's theory is more valid.[1]

To account for the manifest existence of cooperation and related group behavior, such as altruism and restraint in competition, evolutionary theory has recently acquired two kinds of extension. These extensions are, broadly, genetical kinship theory and reciprocity theory. Most of the recent activity, both in fieldwork and in further developments of theory, has been on the side of kinship. Formal approaches have varied, but kinship theory has increasingly taken a gene's-eye view of natural selection (Dawkins 1976). A gene, in effect, looks beyond its mortal bearer to the potentially immortal set of its replicas existing in other related individuals. If the players are sufficiently closely related, altruism can benefit reproduction of the set, despite losses to the individual altruist. In accord with this theory's predictions, almost all clear cases of altruism, and most observed cooperation—apart from their appearance in the human species—occur in contexts of high relatedness, usually between immediate family members. The evolution of the suicidal barbed sting of the honeybee worker could be taken as paradigm for this line of theory (Hamilton 1972).[2]

Conspicuous examples of cooperation (although almost never of ultimate self-sacrifice) also occur where relatedness is low or absent. Mutually advantageous symbioses offer striking examples such as these: the fungus and alga that compose a lichen; the ants and ant-acacias, where the trees house and feed the ants which, in turn, protect the trees (Janzen 1966); and the fig wasps and fig tree, where wasps, which are parasites of fig flowers, serve as the tree's sole means of pollination and seed set (Wiebes 1976; Janzen 1979). Usually the course of cooperation in such symbioses is smooth, but sometimes the partners show signs of antagonism, either spontaneous or elicited by particular treatments (Caullery 1952).[3] Although kinship may be involved, as will be discussed later, symbioses mainly illustrate the other recent extension of evolutionary theory—the theory of reciprocity.

Cooperation itself has received comparatively little attention from biologists since the pioneer account of Trivers (1971); but an associated issue, concerning restraint in conflict situations, has been developed theoretically. In this connection, a new concept—that of an evolutionarily stable strategy—has been formally developed (Maynard Smith and Price 1973; Maynard Smith and Parker 1976; Dawkins 1976; Parker 1978). Cooperation in the more normal sense has remained clouded by certain difficulties, particularly those concerning initiation of cooperation from a previously asocial state (Elster 1979) and its stable maintenance once established. A formal theory of cooperation is increasingly needed. The renewed emphasis on individualism has focused on the frequent ease of cheating. Such cheating makes the stability of even mutually advantageous symbioses appear more questionable than under the old view of adaptation for species benefit. At the same time, other cases that once appeared firmly in the domain of kinship theory now begin to reveal that the players are not closely enough related for much kinship-based altruism to be expected. This applies both to cooperative breeding in birds (Emlen 1978; Stacey 1979) and to cooperative acts more generally in primate groups (Harcourt 1978; Parker 1978; Wrangham 1979). Either the appearances of cooperation are deceptive—they are cases of part-kin altruism and part cheating—or a larger part of the behavior is attributable to stable reciprocity. Previous accounts that already invoke reciprocity, however, underemphasize the stringency of its conditions (Ligon and Ligon 1978).

The contribution of this chapter to biology is new in three ways:

1. In a biological context, the model is novel in its probabilistic treatment of the possibility that two individuals may interact again. This allows light to be shed on certain specific biological processes such as aging and territoriality.
2. The analysis of the evolution of cooperation considers not only the final stability of a given strategy, but also the initial viability of a strategy in an environment dominated by non-cooperating individuals, as well as the robustness of a strategy in a variegated environment composed of other individuals using a variety of more or less sophisticated strategies. This approach allows a richer understanding of the full chronology of the evolution of cooperation than has previously been possible.
3. The applications include behavioral interaction at the microbial level. This leads to some speculative suggestions of rationales able to account for the existence of both chronic and acute phases in many diseases, and for a certain class of genetic defects, exemplified by Down's syndrome.

Many of the benefits sought by living things are disproportionally available to cooperating groups. While there are considerable differences in what is meant by the terms "benefits" and "sought," this statement, insofar as it is true, lays down a fundamental basis for all social life. The problem is that while an individual can benefit from mutual cooperation, each one can also do even better by exploiting the cooperative efforts of others. Over a period of time, the same individuals may interact again, allowing for complex patterns of strategic interactions. As the earlier chapters have shown, the Prisoner's Dilemma allows a formalization of the strategic possibilities inherent in such situations.[4]

Apart from being the solution in game theory, defection in a single encounter is also the solution in biological evolution.[5] It is the outcome of inevitable evolutionary trends through mutation and natural selection: if the payoffs are in terms of fitness, and the interactions between pairs of individuals are random *and not repeated,* then any population with a mixture of heritable strategies evolves to a state where all individuals are defectors. Moreover, no single differing mutant strategy can do better than others when the population is using this strategy. When the players will never meet again, the strategy of defection is the only stable strategy.

In many biological settings, the same two individuals may meet more than once. If an individual can recognize a previous interactant and remember some aspects of the prior outcomes, then the strategic situation becomes an iterated Prisoner's Dilemma with a much richer set of possibilities. A strategy could use the history of the interaction so far to determine the likelihood of its cooperating or defecting on the current move. But, as previously explained, if there is a *known number* of interactions between a pair of individuals, to defect always is still evolutionarily stable and is still the only strategy which is. The reason is that defection on the last interaction would be optimal for both sides, and consequently so would defection on the next-to-last interaction, and so on back to the first interaction.

The model developed in chapter 1 is based on the more realistic assumption that the number of interactions is not fixed in advance. Instead, there is some probability, *w,* that after the current interaction the same two individuals will meet again.[6] Biological factors that affect the magnitude of this probability of meeting again include the average life-span, relative mobility, and health of the individuals. For any value of *w,* the strategy of unconditional defection (ALL D) is always stable; if everyone is using this strategy, no mutant strategy can successfully invade the population.

Stated formally, a strategy is evolutionarily stable if a population of individuals using that strategy cannot be invaded by a rare mutant adopting a different strategy.[7] There may be many evolutionarily stable strategies. In fact, proposition 1 of the first chapter showed that when *w* is sufficiently great, there is no single best strategy regardless of the behavior of the others in the population. Just because there is no single best strategy, it does not follow that analysis is hopeless. On the contrary, chapters 2 and 3 showed that it is possible to analyze not only the stability of a given strategy, but also its robustness and initial viability.

Surprisingly, there is a broad range of biological reality that is encompassed by this game-theoretic approach. To start with, an organism does not need a brain to employ a strategy. Bacteria, for example, have a basic capacity to play games in that (1) bacteria are highly responsive to selected aspects of their environment, especially their chemical environment; (2) this implies that they can respond differentially to what other organisms around them are doing; (3) these conditional strategies of behavior can certainly be inherited; and (4) the behavior of a bacterium can affect the fitness of other organisms around it, just as the behavior of other organisms can affect the fitness of a bacterium. Recent evidence shows that even a virus can use a conditional strategy (Ptashne, Johnson, and Pabo 1982).

While the strategies can easily include differential responsiveness to recent changes in the environment or to cumulative averages over time, in other ways their range of responsiveness is limited. Bacteria cannot "remember" or "interpret" a complex past sequence of changes, and they probably cannot distinguish alternative origins of adverse or beneficial changes. Some bacteria, for example, produce their own antibiotics, called bacteriocins. These are harmless to bacteria of the producing strain, but are destructive to others. A bacterium might easily have production of its own bacteriocin dependent on the perceived presence of like hostile products in its environment, but it could not aim the toxin produced toward an offending initiator.

As one moves up the evolutionary ladder in neural complexity, game-playing behavior becomes richer. The intelligence of primates, including humans, allows a number of relevant improvements: a more complex memory, more complex processing of information to determine the next action as a function of the interaction so far, a better estimate of the probability of future interaction with the same individual, and a better ability to distinguish between different individuals. The discrimination of others may be among the most important of abilities because it allows one to handle interactions with many individuals without having to treat them all the same, thus making possible the rewarding of cooperation from one individual and the punishing of defection from another.

The model of the iterated Prisoner's Dilemma is much less restricted than it may at first appear. Not only can it apply to interactions between two bacteria or interactions between two primates, but it can also apply to the interactions between a colony of bacteria and, say, a primate serving as a host. There is no assumption that payoffs of the two sides are comparable. Provided that the payoffs to each side satisfy the inequalities that define the Prisoner's Dilemma, as given in chapter 1, the results of the analysis will be applicable.

The model does assume that the choices are made simultaneously and with discrete time intervals. For most analytic purposes, this is equivalent to a continuous interaction over time, with the length of time between moves corresponding to the minimum time between a change in behavior by one side and a response by the other. And while the model treats the choices as simultaneous, it would make little difference if they were treated as sequential.[8]

Turning to the development of the theory, the evolution of cooperation can be conceptualized in terms of three separate questions:

1. *Robustness.* What type of strategy can thrive in a variegated environment composed of others using a wide variety of more or less sophisticated strategies?
2. *Stability.* Under what conditions can such a strategy, once fully established, resist invasion by mutant strategies?
3. *Initial viability.* Even if a strategy is robust and stable, how can it ever get a foothold in an environment which is predominantly noncooperative?

The computer tournament described in chapter 2 showed that TIT FOR TAT's strategy of cooperation based on reciprocity was extremely robust. This simple strategy won both rounds of the tournament, and five of the six major variants of the second round. An ecological analysis found that as less successful rules were displaced, TIT FOR TAT continued to do well with the rules which initially did well. Thus cooperation based on reciprocity can thrive in a variegated environment.

Once a strategy has been adopted by the entire population, the question of evolutionary stability deals with whether it can resist invasion by a mutant strategy. The mathematical results of chapter 3 demonstrated that TIT FOR TAT is in fact evolutionarily stable if and only if the interactions between the individuals have a sufficiently large probability of continuing.

TIT FOR TAT is not the only strategy that can be evolutionarily stable. In fact ALL D is evolutionarily stable no matter what the probability is of interaction continuing. This raises the problem of how an evolutionary trend to cooperative behavior could ever have started in the first place.

Genetic kinship theory suggests a plausible escape from the equilibrium of ALL D. Close relatedness of players permits true altruism—sacrifice of fitness by one individual for the benefit of another. True altruism can evolve when the conditions of cost, benefit, and relatedness yield net gains for the altruism-causing genes that are resident in the related individuals (Fisher 1930; Haldane 1955; Hamilton 1963). Not defecting in a single-move Prisoner's Dilemma is altruism of a kind (the individual is foregoing proceeds that might have been taken); so this kind of behavior can evolve if the two players are sufficiently related (Hamilton

1971; Wade and Breden 1980). In effect, recalculation of the payoffs can be done in such a way that an individual has a part interest in the partner's gain (that is, reckoning payoffs in terms of what is called inclusive fitness). This recalculation can often eliminate the inequalities $T > R$ and $P > S$, in which case cooperation becomes unconditionally favored. Thus it is possible to imagine that the benefits of cooperation in Prisoner's Dilemma-like situations can begin to be harvested by groups of closely related individuals. Obviously, as regards pairs, a parent and its offspring or a pair of siblings would be especially promising, and in fact many examples of cooperation or restraint of selfishness in such pairs are known.

Once the genes for cooperation exist, selection will promote strategies that base cooperative behavior on cues in the environment (Trivers 1971). Such factors as promiscuous fatherhood (R. D. Alexander 1974) and events at ill-defined group margins will always lead to uncertain relatedness among potential players. The recognition of any improved correlates of relatedness and use of these cues to determine cooperative behavior will always permit an advance in inclusive fitness. When a cooperative choice has been made, one cue to relatedness is simply the fact of reciprocation of the cooperation. Thus modifiers for more selfish behavior after a negative response from the other are advantageous whenever the degree of relatedness is low or in doubt. As such, the ability to make one's behavior conditional on the behavior of another individual is acquired, and cooperation can spread into circumstances of less and less relatedness. Finally, when the probability of two individuals meeting each other again is sufficiently high, cooperation based on reciprocity can thrive and be evolutionarily stable in a population with no relatedness at all.

A case of cooperation that fits this scenario, at least on first evidence, has been discovered in the spawning relationships in a sea bass (Fischer 1980; Leigh 1977). These fish have the sexual organs of both the male and the female. They form pairs and roughly may be said to take turns at being the high investment partner (laying eggs) and low investment partner (providing sperm to fertilize eggs). Up to ten spawnings occur in a day and only a few eggs are provided each time. Pairs tend to break up if sex roles are not divided evenly. The system appears to allow the evolution of much economy in the size of testes, but Fischer (1980) has suggested that the testes condition may have evolved when the species was more sparse and inclined to inbreed. Inbreeding would imply relatedness in the pairs and this initially may have promoted cooperation without the need of further relatedness.

Another mechanism that can get cooperation started when virtually everyone is using ALL D was demonstrated in chapter 3. This is clustering. Suppose that a small group of individuals is using a strategy such as TIT FOR TAT and that a certain proportion, p, of the interactions of members of this cluster are with other members of the cluster. If the members of the cluster provide a negligible proportion of the interactions for the other individuals, then the score attained by those using ALL D is still virtually equal to the punishment, P, on each move. Then, as shown in chapter 3, if p and w are large enough, a cluster of TIT FOR TAT individuals can become initially viable in an environment composed overwhelmingly of ALL D.

Clustering is often associated with kinship, and the two mechanisms can reinforce each other in promoting the initial viability of reciprocal cooperation. However, it is possible for clustering to be effective without kinship.

Even without kinship, TIT FOR TAT can intrude in a cluster on a population of ALL D, even though ALL D is evolutionarily stable. This is possible because a cluster of TIT FOR TATs gives each member a nontrivial probability of meeting another individual who will reciprocate the cooperation. While this suggests a mechanism for the initiation of cooperation, it also raises the question about whether the reverse could happen once a strategy like TIT FOR TAT becomes established itself. Actually proposition 7 of chapter 3 demonstrates that there is an interesting asymmetry here: the gear wheels of social evolution have a ratchet.

The chronological story that emerges from this analysis is the following. ALL D is the primeval state and is evolutionarily stable. But cooperation based on reciprocity can gain a foothold through two different mechanisms. First, there can be kinship between mutant strategies, giving the genes of the mutants some stake in each other's success, thereby altering the payoff of the interaction when viewed from the perspective of the gene rather than the individual. A second mechanism to overcome total defection is for the mutant strategies to arrive in a cluster so that they provide a nontrivial proportion of the interactions each has, even if they are so few as to provide a negligible proportion of the interactions which the ALL D individuals have. Then the tournament approach described in chapter 2 demonstrates that once a variety of strategies is present, TIT FOR TAT is an extremely robust one. It does well in a wide range of circumstances and gradually displaces all other strategies in an ecological simulation that contains a great variety of more or less sophisticated decision rules. And if the probability that interaction between two individuals will continue is great enough, then TIT FOR TAT is itself evolutionarily stable. Moreover, its stability is especially secure because it can resist the intrusion of whole clusters of mutant strategies. Thus cooperation based on reciprocity can get started in a predominantly noncooperative world, can thrive in a variegated environment, and can defend itself once fully established.

A variety of specific biological applications of this approach follows from two of the requirements for the evolution of cooperation. The basic idea is that an individual must not be able to get away with defecting without the other individuals being able to retaliate effectively. The response requires that the defecting individual not be lost in a sea of anonymous others. Higher organisms avoid this problem by their well-developed ability to recognize many different individuals of their species, but lower organisms must rely on mechanisms that drastically limit the number of different individuals or colonies with which they can interact effectively. The other important requirement to make retaliation effective is that the probability, w, of the same two individuals meeting again must be sufficiently high.

When an organism is not able to recognize the individual with which it had a prior interaction, a substitute mechanism is to make sure that all of its interactions are with the same player. This can be done by maintaining continuous contact with the other. This method is applied in most mutualisms, situations of close association of mutual benefit between members of different species. Examples include a hermit crab and its sea-anemone partner, a cicada and the varied colonies of microorganisms housed in its body, or a tree and its mycorrhizal fungi.

Another mechanism for avoiding the need for recognition is to guarantee the uniqueness of the pairing of players by employing a fixed place of meeting. Consider, for example, mutualisms based on cleaning in which a small fish or a crustacean removes and eats parasites from the body (or even from the inside of the mouth) of a larger fish that is its potential predator. These aquatic cleaner mutualisms occur in coastal and reef situations where animals live in fixed home ranges or territories (Trivers 1971). They seem to be unknown in the free-mixing circumstances of the open sea.

Other mutualisms are also characteristic of situations where continued association is likely, and normally they involve quasi-permanent pairing of individuals, or of inbred or asexual stocks, or of individuals with such stocks (Hamilton 1972 and 1978). Conversely, conditions of free-mixing, and transitory pairing conditions where recognition is impossible, are much more likely to result in exploitation—parasitism, disease, and the like. Thus, whereas ant colonies participate in many symbioses and are sometimes largely dependent on them, honeybee colonies—which are much less permanent in place of abode—have no known symbionts but many parasites (E. O. Wilson 1971; Treisman 1980). The small freshwater animal *Chlorohydra viridissima* has a permanent, stable association with green algae that are always naturally found in its tissues and are very difficult to remove. In this species the alga is transmitted to new generations by way of the egg. *Hydra vulgaris* and *H. attentuata* also associate with algae but do not have egg transmission. In these species it is said that "infection is preceded by enfeeblement of the animals and is accompanied by pathological symptoms indicating a definite parasitism by the plant" (Yonge 1934, p. 13).[9] Again, it is seen that impermanence of association tends to destabilize symbiosis.

In species with a limited ability to discriminate between other members of the same species, reciprocal cooperation can be stable with the aid of a mechanism that reduces the amount of discrimination necessary. Territoriality can serve this purpose. The phrase "stable territories" means that there are two quite different kinds of interaction: with those in neighboring territories where the probability of interaction is high, and with strangers whose probability of future interaction is low. In the case of male territorial birds, songs are used to allow neighbors to recognize each other. Consistent with the theory, such male territorial birds show much more aggressive reactions when the song of an unfamiliar male rather than a neighbor is reproduced nearby (E. O. Wilson 1975, p. 273).

Reciprocal cooperation can be stable with a larger range of individuals if discrimination can cover a wide variety of others with less reliance on supplementary cues such as location. In humans this ability is well developed, and is largely based on the recognition of faces. The extent to which this function has become specialized is revealed by a brain disorder called prosopagnosia. A normal person can name someone from facial features alone, even if the features have changed substantially over the years. People with prosopagnosia are not able to make this association, but have few other neurological symptoms other than a loss of some part of the visual field. The lesions responsible for the disorder occur in an identifiable part of the brain: the underside of both occipital lobes, extending forward to the inner surface of the temporal lobes. This localization of cause, and specificity of effect, indicates that the recognition of individual faces has been an important enough task for a significant portion of the brain's resources to be devoted to it (Geschwind 1979).

Just as the ability to recognize the other player is invaluable in extending the range of stable cooperation, the ability to monitor cues for the likelihood of continued interaction is helpful as an indication of when reciprocal cooperation is or is not stable. In particular, when the relative importance of future interactions, *w*, falls below the threshold for stability, it will no longer pay to reciprocate the other's cooperation.[10] Illness in one partner leading to reduced viability would be one detectable sign of declining *w*. Both animals in a partnership would then be expected to become less cooperative. Aging of a partner would be very like disease in this respect, resulting in an incentive to defect so as to take a one-time gain when the probability of future interaction becomes small enough.

These mechanisms could operate even at the microbial level. Any symbiont that still has a chance to spread to other hosts by some process of infection would be expected to shift from mutualism to parasitism when the probability of continued interaction with the original host lessened. In the more parasitic phase, it could exploit the host more severely by producing more of the forms able to disperse and infect. This phase would be expected when the host is severely injured, has contracted some other wholly parasitic infection that threatens death, or when it manifests signs of age. In fact, bacteria that are normal and seemingly harmless or even beneficial in the gut can be found contributing to sepsis in the body when the gut is perforated, implying a severe wound (Savage 1977). And normal inhabitants of the body surface (like *Candida albicans*) can become invasive and dangerous in either sick or elderly persons.

It is possible also that this argument has some bearing on the causes of cancer, insofar as it turns out to be due to viruses potentially latent in the genome (Manning 1975; Orlove 1977). Cancers do tend to have their onset at ages when the chances of transmission from one generation to the next are rapidly declining (Hamilton 1966). One tumor-causing virus, that of Burkitt's lymphoma, may have alternatives of slow or fast production of infectious stages. The slow form appears as a chronic mononucleosis, the fast as an acute mononucleosis or as a lymphoma (Henle, Henle, and Lenette 1979). The point of interest is that, as some evidence suggests, lymphoma can be triggered by the host's contracting malaria. The lymphoma grows extremely fast and so can probably compete with malaria for transmission (possibly by mosquitoes) before death results. Considering other cases of simultaneous infection by two or more species of pathogen, or by two strains of the same one, the present theory may have relevance more generally to whether a disease will follow a slow, jointly optimal exploitation course ("chronic" for the host) or a rapid severe exploitation ("acute" for the host). With single infection the slow course would be expected. With double infection, crash exploitation might—as dictated by implied payoff functions—begin immediately, or have onset later at an appropriate stage of aging.[11]

The model of the iterated Prisoner's Dilemma could also be tentatively applied to the increase with maternal age of certain kinds of genetic defects (Stern 1973). This effect leads to various conditions of severely handicapped offspring, Down's syndrome (caused by an extra copy of chromosome 21) being the most familiar example. It depends almost entirely on failure of the normal separation of the paired chromosomes in the mother, and this suggests the possible connection with the theory. Cell divisions during formation of the ovum (but usually not sperm) are characteristically asymmetrical, with rejection (as a so-called polar body) of chromosomes that go to the unlucky pole of the cell. It seems possible that, while homologous chromosomes generally stand to gain by steadily cooperating in a diploid organism, the situation is a Prisoner's Dilemma: a chromosome which can be "first to defect" can get itself into the egg nucleus rather than the polar body. One may hypothesize that such an action triggers similar attempts by the homologue in subsequent divisions, and when both members of a homologous pair try it at once, an extra chromosome in the offspring could be the occasional result. The fitness of the bearers of extra chromosomes is generally extremely low, but a chromosome that lets itself be sent to the polar body makes a fitness contribution of zero. Thus P is greater than S. For the model to work, an incident of "defection" in one developing egg would have to be perceptible by others still waiting. That this triggering action would occur is pure speculation, as is the feasibility of self-promoting behavior by chromosomes during such a cell division. But the effects do not seem inconceivable: a bacterium, after all, with its single chromosome, can do complex conditional things. Given such effects, the model would explain the much greater incidence of abnormal chromosome increase in eggs (and not sperm) with parental age.

In this chapter Darwin's emphasis on individual advantage has been formalized in terms of game theory. This formulation establishes conditions under which cooperation in biological systems based on reciprocity can evolve even without foresight by the participants.

IV

Advice for Participants and Reformers

CHAPTER 6

How to Choose Effectively

WHILE FORESIGHT is not necessary for the evolution of cooperation, it can certainly be helpful. Therefore, this chapter and the next are devoted to offering advice to participants and reformers respectively.

This chapter offers advice to someone who is in a Prisoner's Dilemma. From an individual's point of view, the object is to score as well as possible over a series of interactions with another player who is also trying to score well. Since the game is a Prisoner's Dilemma, the player has a short-run incentive to defect, but can do better in the long run by developing a pattern of mutual cooperation with the other. The analysis of the Computer Tournament and the results of the theoretical investigations provide some useful information about what strategies are likely to work under different conditions, and why. The purpose of this chapter is to translate these findings into advice for a player.

The advice takes the form of four simple suggestions for how to do well in a durable iterated Prisoner's Dilemma:

1. Don't be envious.
2. Don't be the first to defect.
3. Reciprocate both cooperation and defection.
4. Don't be too clever.

1. Don't be envious

People are used to thinking about zero-sum interactions. In these settings, whatever one person wins, another loses. A good example is a chess tournament. In order to do well, the contestant must do better than the other player in the game most of the time. A win for White is necessarily a loss for Black.

But most of life is *not* zero-sum. Generally, both sides can do well, or both can do poorly. Mutual cooperation is often possible, but not always achieved. That is why the Prisoner's Dilemma is such a useful model for a wide variety of everyday situations.

In my classes, I have often had pairs of students play the Prisoner's Dilemma for several dozen moves. I tell them that the object is to score well for themselves, as if they were getting a dollar a point. I also tell them that it should not matter to them whether they score a little better or a little worse than the other player, so long as they can collect as many "dollars" for themselves as possible.

These instructions simply do not work. The students look for a standard of comparison to see if they are doing well or poorly. The standard, which is readily available to them, is the comparison of their score with the score of the other player. Sooner or later, one student defects to get ahead, or at least to see what will happen. Then the other usually defects so as not to get behind. Then the situation is likely to deteriorate with mutual recriminations. Soon the players realize that they are not doing as well as they might have, and one of them tries to restore mutual cooperation. But the other is not sure whether this is a ploy that will lead to being exploited again as soon as cooperation begins once more.

People tend to resort to the standard of comparison that they have available—and this standard is often the success of the other player relative to their own success.[1] This standard leads to envy. And envy leads to attempts to rectify any advantage the other player has attained. In this form of Prisoner's Dilemma, rectification of the other's advantage can only be done by defection. But defection leads to more defection and to mutual punishment. So envy is self-destructive.

Asking how well you are doing compared to how well the other player is doing is not a good standard unless your goal is to destroy the other player. In most situations, such a goal is impossible to achieve, or likely to lead to such costly conflict as to be very dangerous to pursue. When you are not trying to destroy the other player, comparing your score to the other's score simply risks the development of self-destructive envy. A better standard of comparison is how well you are doing relative to how well someone else could be doing in your shoes. Given the strategy of the other player, are you doing as well as possible? Could someone else in your situation have done better with this other player? This is the proper test of successful performance.[2]

TIT FOR TAT won the tournament because it did well in its interactions with a wide variety of other strategies. On average, it did better than any other rule with the other strategies in the tournament. Yet TIT FOR TAT never once scored better in a game than the other player! In fact, it can't. It lets the other player defect first, and it never defects more times than the other player has defected. Therefore, TIT FOR TAT achieves either the same score as the other player, or a little less. TIT FOR TAT won the tournament, not by beating the other player, but by eliciting behavior from the other player which allowed both to do well. TIT FOR TAT was so consistent at eliciting mutually rewarding outcomes that it attained a higher overall score than any other strategy.

So in a non-zero-sum world you do not have to do better than the other player to do well for yourself. This is especially true when you are interacting with many different players. Letting each of them do the same or a little better than you is fine, as long as you tend to do well yourself. There is no point in being envious of the success of the other player, since in an iterated Prisoner's Dilemma of long duration the other's success is virtually a prerequisite of your doing well for yourself.

Congress provides a good example. Members of Congress can cooperate with each other without providing threats to each other's standing at home. The main threat to a legislator is not the relative success of another legislator from another part of the country, but from someone who might mount a challenge in the home district. Thus there is not much point in begrudging a fellow legislator the success that comes from mutual cooperation.

Likewise in business. A firm that buys from a supplier can expect that a successful relationship will earn profit for the supplier as well as the buyer. There is no point in being envious of the supplier's profit. Any attempt to reduce it through an uncooperative practice, such as by not paying your bills on time, will only encourage the supplier to take retaliatory action. Retaliatory action could take many forms, often without being explicitly labeled as punishment. It could be less prompt deliveries, lower quality control, less forthcoming attitudes on volume discounts, or less timely news of anticipated changes in market conditions (Macaulay 1963). The retaliation could make the envy quite expensive. Instead of worrying about the relative profits of the seller, the buyer should consider whether another buying strategy would be better.

2. Don't be the first to defect

Both the tournament and the theoretical results show that it pays to cooperate as long as the other player is cooperating.

The tournament results from chapter 2 are very striking. The single best predictor of how well a rule performed was whether or not it was nice, which is to say, whether or not it would ever be the first to defect. In the first round, each of the top eight rules were nice, and not one of the bottom seven were nice. In the second round, all but one of the top fifteen rules were nice (and that one ranked eighth). Of the bottom fifteen rules, all but one were not nice.

Some of the rules that were not nice tried quite sophisticated methods of seeing what they could get away with. For example, TESTER tried an initial defection and then promptly backed off if the other player retaliated. As another example, TRANQUILIZER tended to wait a dozen or two moves before defecting to see if the other player would let itself be lulled and occasionally exploited. If so, TRANQUILIZER threw in additional defections at more frequent intervals, until it was forced to back off by the other's response. But neither of these strategies which experimented with being the first to defect did particularly well. There were too many other players who were not exploitable by virtue of their willingness to retaliate. The resulting conflicts were sometimes quite costly.

Even many of the experts did not appreciate the value of avoiding unnecessary conflict by being nice. In the first round, almost half of the entries by game theorists were not nice. And in the second round, which could take into account the very clear results of the first round, about a third of the entries tried strategies that were not nice. But to little avail.

The theoretical results of chapter 3 provide another way of looking at why nice rules do so well. A population of nice rules is the hardest type to invade because nice rules do so well with each other. Furthermore, a population of nice rules which can resist the invasion of a single mutant rule can resist the invasion of any cluster of other rules (see page 67).

The theoretical results provide an important qualification to the advantages of using a nice strategy. When the future of the interaction is not important enough relative to immediate gains from defection, then simply waiting for the other to defect is not such a good idea. It is important to bear in mind that TIT FOR TAT is a stable strategy only when the discount parameter, *w*, is high enough relative to the payoff parameters, *R*, *S*, *T*, and *P*. In particular, proposition 2 shows that if the discount parameter is not high enough and the other player is using TIT FOR TAT, a player is better off alternating defection and cooperation, or even always defecting. Therefore, if the other player is not likely to be seen again, defecting right away is better than being nice.

This fact has unfortunate implications for groups who are known to move from one place to another. An anthropologist finds that a Gypsy approaches a non-Gypsy expecting trouble, and a non-Gypsy approaches a Gypsy suspiciously, expecting double-dealing.

> For example, a physician was called in to attend a very sick Gypsy baby; he was not the first doctor called, but he was the first willing to come. We escorted him toward the back bedroom, but he stopped short of the threshold of the patient's room. "This visit will be fifteen dollars, and you owe me five dollars from the last time. Pay me the twenty dollars before I see the patient," he demanded. "Okay, okay, you'll get it—just look at the baby now," the Gypsies pleaded. Several more go-arounds occurred before I intervened. Ten dollars changed hands and the doctor examined the patient. After the visit, I discovered the Gypsies, in revenge, did not intend to pay the other ten dollars. (Gropper 1975, pp. 106–7)

In a California community, Gypsies were again found not to pay all of a doctor's bill, but municipal fines *were* paid promptly (Sutherland 1975, p. 70). These fines were usually for breaking garbage regulations. This was among a group of Gypsies who returned to the same town every winter. Presumably, the Gypsies knew that they had an ongoing relationship with the garbage collection service of that town, and could not shop around for another service. Conversely, there were always enough doctors in the area for them to break off one relationship and start another when necessary.[3]

Short interactions are not the only condition which would make it pay to be the first to defect. The other possibility is that cooperation will simply not be reciprocated. If everyone else is using a strategy of always defecting, then a single individual can do no better than to use this same strategy. But, as shown in chapter 3, if even a small proportion of one's interactions are going to be with others who are using a responsive strategy like TIT FOR TAT, then it can pay to use TIT FOR TAT rather than to simply defect all the time like most of those in the population. In the numerical example presented there, it took only 5 percent of one's interactions to be with like-minded TIT FOR TAT players to make the members of this small cluster do better than the typical defecting member of the population.[4]

Will there be anyone out there to reciprocate one's own initial cooperation? In some circumstances this will be hard to tell in advance. But if there has been enough time for many different strategies to be tried, and for some way of making the more successful strategies become more common, then one can be fairly confident that there will be individuals out there who will reciprocate cooperation. The reason is that even a relatively small cluster of discriminating nice rules can invade a population of meanies, and then thrive on their good scores with each other. And once nice rules get a foothold they can protect themselves from reinvasion by meanies.

Of course, one could try to "play it safe" by defecting until the other player cooperates, and only then starting to cooperate. The tournament results show, however, that this is actually a very risky strategy. The reason is that your own initial defection is likely to set off a retaliation by the other player. This will put the two of you in the difficult position of trying to extricate yourselves from an initial pattern of exploitation or mutual defection. If you punish the other's retaliation, the problem can echo into the future. And if you forgive the other, you risk appearing to be exploitable. Even if you can avoid these long-term problems, a prompt retaliation against your initial defection can make you wish that you had been nice from the start.

The ecological analysis of the tournament revealed another reason why it is risky to be the first to defect. The only rule that was not nice and that scored among the top fifteen in the second round of the tournament was the eighth-ranking rule, HARRINGTON. This rule did fairly well because it scored well with the lower ranking entries in the tournament. In hypothetical future rounds of the tournament, the lower ranking entries became a smaller and smaller proportion of the population. Eventually, the non-nice rule that originally scored well had fewer and fewer strategies it could do well with. Then it too suffered and eventually died out. Thus the ecological analysis shows that doing well with rules that do not score well themselves is eventually a self-defeating process. The lesson is that not being nice may look promising at first, but in the long run it can destroy the very environment it needs for its own success.

3. Reciprocate both cooperation and defection

The extraordinary success of TIT FOR TAT leads to some simple, but powerful advice: practice reciprocity. After cooperating on the first move, TIT FOR TAT simply reciprocates whatever the other player did on the previous move. This simple rule is amazingly robust. It won the first round of the Computer Tournament for the Prisoner's Dilemma by attaining a higher average score than any other entry submitted by professional game theorists. And when this result was publicized for the contestants in the second round, TIT FOR TAT won again. The victory was obviously a surprise, since anyone could have submitted it to the second round after seeing its success in the first round. But obviously people hoped they could do better—and they were wrong.

TIT FOR TAT not only won the tournament itself, but did better than any other rule in hypothetical future rounds. This indicates that TIT FOR TAT not only does well with the original great variety of rules, but also does well with successful rules which would be likely to show up in the future in greater proportions. It does not destroy the basis of its own success. On the contrary, it thrives on interactions with other successful rules.

The reciprocity embodied in TIT FOR TAT is good for theoretical reasons as well. When the future is important enough relative to the present, TIT FOR TAT is collectively stable. This means that if everyone is using TIT FOR TAT, there is no better advice to offer a particular player than to use TIT FOR TAT as well. Or putting it another way, if you are sure the other player is using TIT FOR TAT and the interaction will last long enough, then you might as well do the same. But the beauty of the reciprocity of TIT FOR TAT is that it is good in such a wide range of circumstances.

In fact, TIT FOR TAT is very good at discriminating between rules which will return its own initial cooperation and those which will not. It is even maximally discriminating in the sense introduced in chapter 3 (see page 66). This allows it to invade a world of meanies in the smallest possible cluster, as demonstrated in proposition 6. Moreover, it will reciprocate a defection as well as a cooperation, making it provocable. And proposition 4 demonstrates that being provocable is actually required for a nice rule like TIT FOR TAT to resist invasion.

In responding to a defection from the other player, TIT FOR TAT represents a balance between punishing and being forgiving. TIT FOR TAT always defects exactly once after each defection by the other, and TIT FOR TAT was very successful in the tournament. This suggests the question of whether always doing exactly one-for-one is the most effective balance. It is hard to say because rules with slightly different balances were not submitted. What is clear is that extracting more than one defection for each defection of the other risks escalation. On the other hand, extracting less than one-for-one risks exploitation.

TIT FOR TWO TATS is the rule that defects only if the other player has defected in both of the previous two moves. Therefore it returns one-for-two. This relatively forgiving rule would have won the first round of the Computer Tournament for the Prisoner's Dilemma had it been submitted. It would have done so well because it would have avoided mutual recriminations with some of the other rules that caused trouble even for TIT FOR TAT. Yet in the second round of the tournament, when TIT FOR

TWO TATS was actually submitted, it did not even score in the top third. The reason is that the second round contained some rules that were able to exploit its willingness to forgive isolated defections.

The moral of the story is that the precise level of forgiveness that is optimal depends upon the environment. In particular, if the main danger is unending mutual recriminations, then a generous level of forgiveness is appropriate. But, if the main danger is from strategies that are good at exploiting easygoing rules, then an excess of forgiveness is costly. While the exact balance will be hard to determine in a given environment, the evidence of the tournament suggests that something approaching a one-for-one response to defection is likely to be quite effective in a wide range of settings. Therefore it is good advice to a player to reciprocate defection as well as cooperation.

4. *Don't be too clever*

The tournament results show that in a Prisoner's Dilemma situation it is easy to be *too* clever. The very sophisticated rules did not do better than the simple ones. In fact, the so-called maximizing rules often did poorly because they got into a rut of mutual defection. A common problem with these rules is that they used complex methods of making inferences about the other player—and these inferences were wrong. Part of the problem was that a trial defection by the other player was often taken to imply that the other player could not be enticed into cooperation. But the heart of the problem was that these maximizing rules did not take into account that their *own* behavior would lead the other player to change.

In deciding whether to carry an umbrella, we do not have to worry that the clouds will take our behavior into account. We can do a calculation about the chance of rain based on past experience. Likewise in a zero-sum game, such as chess, we can safely use the assumption that the other player will pick the most dangerous move that can be found, and we can act accordingly. Therefore it pays for us to be as sophisticated and as complex in our analysis as we can.

Non-zero-sum games, such as the Prisoner's Dilemma, are not like this. Unlike the clouds, the other player can respond to your own choices. And unlike the chess opponent, the other player in a Prisoner's Dilemma should not be regarded as someone who is out to defeat you. The other player will be watching your behavior for signs of whether you will reciprocate cooperation or not, and therefore your own behavior is likely to be echoed back to you.

Rules that try to maximize their own score while treating the other player as a fixed part of the environment ignore this aspect of the interaction, no matter how clever they are in calculating under their limiting assumptions. Therefore, it does not pay to be clever in modeling the other player if you leave out the reverberating process in which the other player is adapting to you, you are adapting to the other, and then the other is adapting to your adaptation and so on. This is a difficult road to follow with much hope for success. Certainly none of the more or less complex rules submitted in either round of the tournament was very good at it.

Another way of being too clever is to use a strategy of "permanent retaliation." This is the strategy of cooperating as long as the other player cooperates, but then never again cooperating after a single defection by the other. Since this strategy is nice, it does well with the other nice rules. And it does well with rules which were not very responsive, such as the completely random rule. But with many others it does poorly because it gives up too soon on rules that try an occasional defection, but are ready to back off once punished. Permanent retaliation may seem clever because it provides the maximum incentive to avoid defection. But it is too harsh for its own good.

There is yet a third way in which some of the tournament rules are too clever: they employ a probabilistic strategy that is so complex that it cannot be distinguished by the other strategies from a purely random choice. In other words, too much complexity can appear to be total chaos. If you are using a strategy which appears random, then you also appear unresponsive to the other player. If you are unresponsive, then the other player has no incentive to cooperate with you. So being so complex as to be incomprehensible is very dangerous.

Of course, in many human situations a person using a complex rule can explain the reasons for each choice to the other player. Nevertheless, the same problem arises. The other player may be dubious about the reasons offered when they are so complicated that they appear to be made up especially for that occasion. In such circumstances, the other player may well doubt that there is any responsiveness worth fostering. The other player may thus regard a rule that appears to be unpredictable as unreformable. This conclusion will naturally lead to defection.

One way to account for TIT FOR TAT's great success in the tournament is that it has great clarity: it is eminently comprehensible to the other player. When you are using

TIT FOR TAT, the other player has an excellent chance of understanding what you are doing. Your one-for-one response to any defection is an easy pattern to appreciate. Your future behavior can then be predicted. Once this happens, the other player can easily see that the best way to deal with TIT FOR TAT is to cooperate with it. Assuming that the game is sufficiently likely to continue for at least one more interaction, there is no better plan when meeting a TIT FOR TAT strategy than to cooperate now so that you will be the recipient of a cooperation on the very next move.

Once again, there is an important contrast between a zero-sum game like chess and a non-zero-sum game like the iterated Prisoner's Dilemma. In chess, it is useful to keep the other player guessing about your intentions. The more the other player is in doubt, the less efficient will be his or her strategy. Keeping one's intentions hidden is useful in a zero-sum setting where any inefficiency in the other player's behavior will be to your benefit. But in a non-zero-sum setting it does not always pay to be so clever. In the iterated Prisoner's Dilemma, you benefit from the other player's cooperation. The trick is to encourage that cooperation. A good way to do it is to make it clear that you will reciprocate. Words can help here, but as everyone knows, actions speak louder than words. That is why the easily understood actions of TIT FOR TAT are so effective.

CHAPTER 7

How to Promote Cooperation

THIS CHAPTER takes the perspective of a reformer. It asks how the strategic setting itself can be transformed in order to promote cooperation among the players. The previous chapter took a different perspective. There the problem was how to advise an individual who was in a *given* environment. If the strategic setting allowed long enough interactions between individuals, much of the advice pointed to reasons why an egoist should be willing to cooperate even though there is a short-term incentive not to cooperate. But if the interaction was not very durable, then an egoist would be better off going for short-run benefits, and defecting. This chapter, on the other hand, does not take the strategic setting as given. Instead it asks how one can promote cooperation by transforming the strategic setting itself—for example, by enlarging the shadow of the future.

Usually one thinks of cooperation as a good thing. This is the natural approach when one takes the perspective of the players themselves. After all, mutual cooperation is good for both players in a Prisoner's Dilemma. So this chapter will be written from the point of view of how to promote cooperation. Yet, as previously suggested, there are situations in which one wants to do just the opposite. To prevent businesses from fixing prices, or to prevent potential enemies from coordinating their actions, one would want to turn the approach around and do the opposite of what would promote cooperation.

The Prisoner's Dilemma itself is named for such a situation. The original story is that two accomplices to a crime are arrested and questioned separately. Either can defect against the other by confessing and hoping for a lighter sentence. But if both confess, their confessions are not as valuable. On the other hand, if both cooperate with each other by refusing to confess, the district attorney can only convict them on a minor charge. Assuming that neither player has moral qualms about, or fear of, squealing, the payoffs can form a Prisoner's Dilemma (Luce and Raiffa 1957, pp. 94–95). From society's point of view, it is a good thing that the two accomplices have little likelihood of being caught in the same situation again soon, because that is precisely the reason why it is to each of their individual advantages to double-cross the other.

As long as the interaction is not iterated, cooperation is very difficult. That is why an important way to promote cooperation is to arrange that the same two individuals will meet each other again, be able to recognize each other from the past, and to recall how the other has behaved until now. This continuing interaction is what makes it possible for cooperation based on reciprocity to be stable.

The advice dealing with how this mutual cooperation can be promoted comes in three categories: making the future more important relative to the present; changing the payoffs to the players of the four possible outcomes of a move; and teaching the players values, facts, and skills that will promote cooperation.

1. Enlarge the shadow of the future

Mutual cooperation can be stable if the future is sufficiently important relative to the present. This is because the players can each use an implicit threat of retaliation against the other's defection—if the interaction will last long enough to make the threat effective. Seeing how this works in a numerical example will allow the formulation of the alternative methods that can enlarge the shadow of the future.

As previously, suppose that a payoff received in the next move is worth only some fixed percentage of the same payoff received in the current move. Recall that this discount parameter, w, reflects two reasons why the future is typically less important than the present. In the first place, the interaction may not continue. One or the other player may die, go bankrupt, move away, or the relationship may end for any other reason. Since these factors cannot be predicted with certainty, the next move is not as important as the current one. There may be no next move. A second reason that the future is less important than the present is that individuals typically prefer to get a given benefit today, rather than having to wait for the same benefit until tomorrow. Both of these effects combine to make the next move less important than the present one.

The numerical example is the familiar one of an iterated Prisoner's Dilemma with the payoffs as follows: the temptation to defect while the other is cooperating gives $T=5$, the reward for mutual cooperation is $R=3$, the punishment for mutual defection is $P=1$, and the sucker's payoff for cooperating when the other defects is $S=0$. Suppose for a moment that the next move is worth 90 percent of the current move, making $w=.9$. Then if the other player is using TIT FOR TAT, it does not pay for you to defect. This follows directly from proposition 2, which tells when TIT FOR TAT is collectively stable, but it can be calculated again to see how it works. Never defecting when meeting a TIT FOR TAT strategy will give a score of R on each move. After taking account of the discount rate, this is accumulated into a total expected score of $R + wR + w^2R$ . . . which is $R/(1-w)$. For $R=3$ and $w=.9$ this is 30 points.

You can't do better. If you always defect, you get the tempting payoff, $T=5$, on the first move, but thereafter you get only the punishment for mutual defection, $P=1$. This accumulates to 14 points.[1] And 14 points is not as good as the 30 points you could have gotten by cooperating. You could also try alternating defection and cooperation, repeatedly setting TIT FOR TAT up for exploitation, at the cost of being exploited yourself on the alternative moves. This would give 26.3 points.[2] This is better than the 14 points from always defecting, but not as good as the 30 points from always cooperating with TIT FOR TAT. And an implication of proposition 2 is that if these two strategies are not better with TIT FOR TAT than mutual cooperation is, then no other strategy will be better either. When the future casts a large shadow as reflected in the high discount parameter of 90 percent, then it pays to cooperate with someone using TIT FOR TAT. And because of this, it pays to use TIT FOR TAT. And therefore with a large shadow, cooperation based on reciprocity is stable.

The situation changes when the shadow of the future is not so great. To see this, suppose the discount parameter were changed from 90 percent to 30 percent. This reduction might be due to a greater likelihood that the interaction will end soon, or to a greater preference for immediate benefits over delayed gratification, or to any combination of these two factors. Again, suppose that the other player is using TIT FOR TAT. If you cooperate, you will get R per move, as before. Your expected score will be $R/(1-w)$ as before but now this is worth only 4.3 points because of the lower value of w. Can you do better? If you always defect, you get $T=5$ on the first move, and thereafter you get $P=1$. This accumulates to 5.4 points, which is better than you could have gotten by being nice. Alternating defection and cooperation does even better, giving 6.2 points. So as the shadow of the future becomes smaller, it stops paying to be cooperative with another player—even if the other player will reciprocate your cooperation.

And if it does not pay for you to cooperate, it does not pay for the other player to cooperate either. So when the discount parameter is not high enough, cooperation is likely to be missing altogether or to disappear fairly quickly. This conclusion does not depend on the use of TIT FOR TAT, because proposition 3 in chapter 3 [p. 61] showed that *any* strategy that may be the first to cooperate is stable only when the discount parameter is high enough; this means that *no* form of cooperation is stable when the future is not important enough relative to the present.

This conclusion emphasizes the importance of the first method of promoting cooperation: enlarging the shadow of the future. There are two basic ways of doing this: by making the interactions more durable, and by making them more frequent.

The most direct way to encourage cooperation is to make the interactions more durable. For example, a wedding is a public act designed to celebrate and promote the durability of a relationship. Durability of an interaction can help not only lovers, but enemies. The most striking illustration of this point was the way the live-and-let-live system developed during the trench warfare of World War I. As seen in chapter 4, what was unusual about trench warfare was that the same small units of troops would be in contact with each other for extended periods of time. They knew that their interactions would continue because no one was going anywhere. In more mobile wars, a small unit would meet a different enemy unit every time there would be an engagement; consequently it would not pay to initiate cooperation on the hope that the other individual or small unit will reciprocate later. But in static combat, the interaction between two small units is prolonged over a substantial period of time. This prolonged interaction allows patterns of cooperation which are based on reciprocity to be worth trying and allows them to become established.

Another way to enlarge the shadow of the future is to make the interactions more frequent. In such a case the next interaction occurs sooner, and hence the next move looms larger than it otherwise would. This increased rate of interaction would therefore be reflected in an increase in *w*, the importance of the next move relative to the current move.

It is important to appreciate that the discount parameter, *w*, is based on the relative importance of one move and the next, not one time period and the next. Therefore, if the players regard a payoff two years from now as worth only half as much as an equal payoff today, one way to promote cooperation would be to make their interactions more frequent.

A good way to increase the frequency of interactions between two given individuals is to keep others away. For example, when birds establish a territory it means that they will have only a few neighbors. This, in turn, means that they will have relatively frequent interactions with these nearby individuals. The same could be true for a business firm that had a territorial base and bought and sold mainly with only a few firms in its own territory. Likewise, any form of specialization tending to restrict interactions to only a few others would tend to make the interactions with those few more frequent. This is one reason why cooperation emerges more readily in small towns than in large cities. It is also a good reason why firms in a congenial industry try to keep out new firms that might upset the cozy restraints on competition that have grown up in the restricted industry. Finally, an itinerant trader or day worker will have an easier time developing cooperative relationships with customers if the customers see the worker on a regular basis rather than only at long and unpredictable intervals. The principle is always the same: frequent interactions help promote stable cooperation.

Hierarchy and organization are especially effective at concentrating the interactions between specific individuals. A bureaucracy is structured so that people specialize, and so that people working on related tasks are grouped together. This organizational practice increases the frequency of interactions, making it easier for workers to develop stable cooperative relationships. Moreover, when an issue requires coordination between different branches of the organization, the hierarchical structure allows the issue to be referred to policy makers at higher levels who frequently deal with each other on just such issues. By binding people together in a long-term, multilevel game, organizations increase the number and importance of future interactions, and thereby promote the emergence of cooperation among groups too large to interact individually. This in turn leads to the evolution of organizations for the handling of larger and more complex issues.

Concentrating the interactions so that each individual meets often with only a few others has another benefit besides making cooperation more stable. It also helps get cooperation going. As mentioned in the discussion of clustering in chapter 3, even a small cluster of individuals can invade a large population of meanies. The members of the cluster must have a nontrivial proportion of their interactions with each other, even though the majority of their interactions may be with the general population. The numerical example showed how easy it was for a small cluster of TIT FOR TAT players to invade a population of players who always defect. With the standard illustrative payoff values ($T=5$, $R=3$, $P=1$, and $S=0$) and a moderate discount parameter ($w=.9$), members of the cluster needed just 5 percent of their interactions to be with other members of the cluster in order for the cooperation to get started in a mean world.

Concentrating the interactions is one way to make two individuals meet more often. In a bargaining context, another way to make their interactions more frequent is to break down the issues into small pieces. An arms control or disarmament treaty, for example, can be broken down into many stages. This would allow the two parties to make many relatively small moves rather than one or two large moves. Doing it this way makes reciprocity more effective. If both sides can know that an inadequate move by the other can be met with a reciprocal defection in the next stage, then both can be more confident that the process will work out as anticipated. Of course, a major question in arms control is whether each side can, in fact, know what the other side actually did on the previous move—whether they cooperated by fulfilling their obligations or defected by cheating. But for any given degree of confidence in each side's ability to detect cheating, having many small steps will help promote cooperation as compared to having just a few big steps. Decomposing the interaction promotes the stability of cooperation by making the gains from cheating on the current move that much less important relative to the gains from potential mutual cooperation on later moves.

Decomposition is a widely practiced principle. Henry Kissinger arranged for the Israeli disengagement from the Sinai after the 1973 war to proceed in stages that were coordinated with Egyptian moves leading to normal relationships with Israel. Businesses prefer to ask for payment for large orders in phases, as the deliveries are made, rather than to wait for a lump sum at the end. Making sure that defection on the present move is not too tempting relative to the whole future course of the interaction is a good way to promote cooperation. But another way is to alter the payoffs themselves.

2. Change the payoffs

A common reaction of someone caught in a Prisoner's Dilemma is that "there ought to be a law against this sort of thing." In fact, getting out of Prisoner's Dilemmas is one of the primary functions of government: to make sure that when individuals do not have private incentives to cooperate, they will be required to do the socially useful thing anyway. Laws are passed to cause people to pay their taxes, not to steal, and to honor contracts with strangers. Each of these activities could be regarded as a giant Prisoner's Dilemma game with many players. No one wants to pay taxes because the benefits are so diffuse and the costs are so direct. But everyone may be better off if each person has to pay so that each can share the benefits of schools, roads, and other collective goods (Schelling 1973). This is a major part of what Rousseau meant when he said that government's role is to make sure that each citizen "will be forced to be free" (Rousseau 1762/1950, p. 18).

What governments do is to change the effective payoffs. If you avoid paying your taxes, you must face the possibility of being caught and sent to jail. This prospect makes the choice of defection less attractive. Even quasi-governments can enforce their laws by changing the payoffs faced by the players. For example, in the original story of the Prisoner's Dilemma, there were two accomplices arrested and interrogated separately. If they belonged to an organized gang, they could anticipate being punished for squealing. This might lower the payoffs for double-crossing their partner so much that neither would confess—and both would get the relatively light sentence that resulted from the mutual cooperation of their silence.

Large changes in the payoff structure can transform the interaction so that it is no longer even a Prisoner's Dilemma. If the punishment for defection is so great that cooperation is the best choice in the short run, no matter what the other player does, then there is no longer a dilemma. The transformation of payoffs does not have to be quite this drastic to be effective, however. Even a relatively small transformation of the payoffs might help make cooperation based on reciprocity stable, despite the fact that the interaction is still a Prisoner's Dilemma. The reason is that the conditions for stability of cooperation are reflected in the relationship between the discount parameter, *w*, and the four outcome payoffs, *T*, *R*, *S*, and *P*.[3] What is needed is for *w* to be large enough relative to these payoffs. If the payoffs change, the situation may change from one in which cooperation is not stable to one in which it is. So, to promote cooperation through modification of the payoffs, it is not necessary to go so far as to eliminate the tension between the short-run incentive to defect and the longer-run incentive to achieve mutual cooperation. It is only necessary to make the long-term incentive for mutual cooperation greater than the short-term incentive for defection.

3. Teach people to care about each other

An excellent way to promote cooperation in a society is to teach people to care about the welfare of others. Parents and schools devote a tremendous effort to teaching the young to value the happiness of others. In game theory terms, this means that the adults try to shape the values of children so that the preferences of the new citizens will incorporate not only their own individual welfare, but to some degree at least, the welfare of others. Without doubt, a society of such caring people will have an easier time attaining cooperation among its members, even when caught in an iterated Prisoner's Dilemma.

Altruism is a good name to give to the phenomenon of one person's utility being positively affected by another person's welfare.[4] Altruism is thus a motive for action. It should be recognized, however, that certain kinds of behavior that may look generous may actually take place for reasons other than altruism. For example, giving to charity is often done less out of a regard for the unfortunate than for the sake of the social approval it is expected to bring. And in both traditional and modern societies, gift giving is likely to be part of an exchange process. The motive may be more to create an obligation than to improve the welfare of the recipient (Blau 1968).

From the point of view of the genetics of biological evolution, altruism can be sustained among kin. A mother who risks her own life to save several of her offspring can increase the odds that copies of her genes will survive. This is the basis of genetical kinship theory, as discussed in chapter 5.

Altruism among people can also be sustained through socialization. But there is a serious problem. A selfish individual can receive the benefits of another's altruism and not pay the welfare costs of being generous in return. We have all met spoiled brats, people who expect others to be considerate and generous, but who do not think of the needs of anyone but themselves. Such people need to be treated differently than more considerate people, lest we be exploited by them. This reasoning suggests that the costs of altruism can be controlled by being altruistic to everyone at first, and thereafter only to those who show similar feelings. But this quickly takes one back to reciprocity as the basis for cooperation.

4. Teach reciprocity

TIT FOR TAT may be an effective strategy for an egoist to use, but is it a moral strategy for a person or a country to follow? The answer depends, of course, on one's standard for morality. Perhaps the most widely accepted moral standard is the Golden Rule: Do unto others as you would have them do unto you. In the context of the Prisoner's Dilemma, the Golden Rule would seem to imply that you should always cooperate, since cooperation is what you want from the other player. This interpretation suggests that the best strategy from the point of view of morality is the strategy of unconditional cooperation rather than TIT FOR TAT.

The problem with this view is that turning the other cheek provides an incentive for the other player to exploit you. Unconditional cooperation can not only hurt you, but it can hurt other innocent bystanders with whom the successful exploiters will interact later. Unconditional cooperation tends to spoil the other player; it leaves a burden on the rest of the community to reform the spoiled player, suggesting that reciprocity is a better foundation for morality than is unconditional cooperation. The Golden Rule *would* advise unconditional cooperation, since what you would really prefer the other player to do is to let you get away with some defections.

Yet, basing a strategy on reciprocity does not seem to be the height of morality either—at least not according to our everyday intuitions. Reciprocity is certainly not a good basis for a morality of aspiration. Yet it is more than just the morality of egoism. It actually helps not only oneself, but others as well. It helps others by making it hard for exploitative strategies to survive. And not only does it help others, but it asks no more for oneself than it is willing to concede to others. A strategy based on reciprocity can allow the other player to get the reward for mutual cooperation, which is the same payoff it gets for itself when both strategies are doing their best.

The insistence on no more than equity is a fundamental property of many rules based upon reciprocity. It is most clearly seen in the performance of TIT FOR TAT in the Prisoner's Dilemma tournaments. TIT FOR TAT won both rounds of the tournament, but it never received more points in any game than the other player! Indeed, it can't possibly score more than the other player in a game because it always lets the other player defect first, and it will never defect more times than the other player does. It won, not by doing better than the other player, but by eliciting cooperation from the other player. In this way, TIT FOR TAT does well by promoting the mutual interest rather than by exploiting the other's weakness. A moral person couldn't do much better.

What gives TIT FOR TAT its slightly unsavory taste is its insistence on an eye for an eye. This is rough justice indeed. But the real issue is whether there are any better alternatives. In situations where people can rely on a central authority to enforce the community standards, there are alternatives. The punishment might fit the crime without having to be as painful as the crime itself was. When there is no central authority to do the enforcement, the players must rely on themselves to give each other the necessary incentives to elicit cooperation rather than defection. In such a case the real question is just what form this enticement should take.

The trouble with TIT FOR TAT is that once a feud gets started, it can continue indefinitely. Indeed, many feuds seem to have just this property. For example, in Albania and the Middle East, a feud between families sometimes goes on for decades as one injury is repaid by another, and each retaliation is the start of the next cycle. The injuries can echo back and forth until the original violation is lost in the distant past (Black-Michaud 1975). This is a serious problem with TIT FOR TAT. A better strategy might be to return only nine-tenths of a tit for a tat. This would help dampen the echoing of conflict and still provide an incentive to the other player not to try any gratuitous defections. It would be a strategy based on reciprocity, but would be a bit more forgiving than TIT FOR TAT. It is still rough justice, but in a world of egoists without central authority, it does have the virtue of promoting not only its own welfare, but the welfare of others as well.

A community using strategies based upon reciprocity can actually police itself. By guaranteeing the punishment of any individual who tries to be less than cooperative, the deviant strategy is made unprofitable. Therefore the deviant will not thrive, and will not provide an attractive model for others to imitate.

This self-policing feature gives you an extra private incentive to teach it to others—even those with whom you will never interact. Naturally, you want to teach reciprocity to those with whom you will interact so that you can build a mutually rewarding relationship. But you also have a private advantage from another person using reciprocity even if you never interact with that person: the other's reciprocity helps to police the entire community by punishing those who try to be exploitive. And this decreases the number of uncooperative individuals you will have to deal with in the future.

So teaching the use of nice strategies based upon reciprocity helps the pupil, helps the community, and can indirectly help the teacher. No wonder that an educational psychologist, upon hearing of the virtues of TIT FOR TAT, recommended teaching reciprocity in the schools (Calfee 1981, p. 38).

5. Improve recognition abilities

The ability to recognize the other player from past interactions, and to remember the relevant features of those interactions, is necessary to sustain cooperation. Without these abilities, a player could not use any form of reciprocity and hence could not encourage the other to cooperate.

In fact, the scope of sustainable cooperation is dependent upon these abilities. This dependence is most clearly seen in the range of biological illustrations developed in chapter 5. Bacteria, for example, are near the bottom of the evolutionary ladder and have limited ability to recognize other organisms. So they must use a shortcut to recognition: an exclusive relationship with just one other player (the host) at a time. In this way, any changes in a bacterium's environment can be attributed to that one player.[5] Birds are more discriminating—they can distinguish among a number of individual neighboring birds by their songs. This ability to discriminate allows them to develop cooperative relationships—or at least avoid conflictful ones—with several other birds. And as discussed in chapter 5, humans have developed their recognition abilities to the extent of having a part of their brains specialized for the recognition of faces. The expanded ability to recognize individuals with whom one has already interacted allows humans to develop a much richer set of cooperative relationships than birds can.

Yet, even in human affairs, limits on the scope of cooperation are often due to the inability to recognize the identity or the actions of the other players. This problem is especially acute for the achievement of effective international control of nuclear weapons. The difficulty here is verification: knowing with an adequate degree of confidence what move the other player has actually made. For example, an agreement to ban all testing of nuclear weapons has until recently been prevented by the technical difficulty of distinguishing explosions from earthquakes—a difficulty that has now been largely overcome (Sykes and Everden 1982).

The ability to recognize defection when it occurs is not the only requirement for successful cooperation to emerge, but it is certainly an important one. Therefore, the scope of sustainable cooperation can be expanded by any improvements in the players' ability to recognize each other from the past, and to be confident about the prior actions that have actually been taken. This chapter has shown that cooperation among people can be promoted by a variety of other techniques as well, which include enlarging the shadow of the future, changing the payoffs, teaching people to care about the welfare of others, and teaching the value of reciprocity. Promoting good outcomes is not just a matter of lecturing the players about the fact that there is more to be gained from mutual cooperation than mutual defection. It is also a matter of shaping the characteristics of the interaction so that over the long run there can be a stable evolution of cooperation.

V

Conclusions

CHAPTER 8

The Social Structure of Cooperation

IN CONSIDERING how the evolution of cooperation could have begun, some social structure was found to be necessary. In particular, it was shown in chapter 3 that a population of meanies who always defect could not be invaded by a single individual using a nice strategy such as TIT FOR TAT. But if the invaders had even a small amount of social structure, things could be different. If they came in a cluster so that they had even a small percentage of their interactions with each other, then they could invade the population of meanies.

This chapter explores the consequences of additional forms of social structure. Four factors are examined which can give rise to interesting types of social structure: labels, reputation, regulation, and territoriality. A label is a fixed characteristic of a player, such as sex or skin color, which can be observed by the other player. It can give rise to stable forms of stereotyping and status hierarchies. The reputation of a player is malleable and comes into being when another player has information about the strategy that the first one has employed with other players. Reputations give rise to a variety of phenomena. including incentives to establish a reputation as a bully, and incentives to deter others from being bullies. Regulation is a relationship between a government and the governed. Governments cannot rule only through deterrence, but must instead achieve the voluntary compliance of the majority of the governed. Therefore regulation gives rise to the problems of just how stringent the rules and the enforcement procedures should be. Finally, territoriality occurs when players interact with their neighbors rather than with just anyone. It can give rise to fascinating patterns of behavior as strategies spread through a population.

Labels, Stereotypes, and Status Hierarchies

People often relate to each other in ways that are influenced by observable features such as sex, age, skin color, and style of dress. These cues allow a player to begin an interaction with a stranger with an expectation that the stranger will behave like others who share these same observable characteristics. In principle, then, these characteristics can allow a player to know something useful about the other player's strategy even before the interaction begins. This happens because the observed characteristics allow an individual to be labeled by others as a member of a group with similar characteristics. This labeling, in turn, allows the inferences about how that individual will behave.

The expectations associated with a given label need not be learned from direct personal experience. The expectations could also be formed by secondhand experience through the process of sharing of anecdotes. The interpretations given to the cues could even be formed through genetics and natural selection, as when a turtle is able to distinguish the sex of another turtle and respond accordingly.

A *label* can be defined as a fixed characteristic of a player that can be observed by other players when the interaction begins.[1] When there are labels, a strategy can determine a choice based not only on the history of the interaction so far, but also upon the label assigned to the other player.

One of the most interesting but disturbing consequences of labels is that they can lead to self-confirming stereotypes. To see how this can happen, suppose that everyone has either a Blue label or a Green label. Further, suppose that both groups are nice to members of their own group and mean to members of the other group. For the sake of concreteness, suppose that members of both groups employ TIT FOR TAT with each other and always defect with members of the other group. And suppose that the discount parameter, w, is high enough to make TIT FOR TAT a collectively stable strategy (in accordance with proposition 2 of chapter 3). Then a single individual, whether Blue or Green, can do no better than to do what everyone else is doing and be nice to one's own type and mean to the other type.

This incentive means that stereotypes can be stable, even when they are not based on any objective differences. The

Blues believe that the Greens are mean, and whenever they meet a Green, they have their beliefs confirmed. The Greens think that only other Greens will reciprocate cooperation, and they have their beliefs confirmed. If you try to break out of the system, you will find that your own payoff falls and your hopes will be dashed. So if you become a deviant, you are likely to return, sooner or later, to the role that is expected of you. If your label says you are Green, others will treat you as a Green, and since it pays for you to act like Greens act, you will be confirming everyone's expectations.

This kind of stereotyping has two unfortunate consequences: one obvious and one more subtle. The obvious consequence is that everyone is doing worse than necessary because mutual cooperation between the groups could have raised everyone's score. A more subtle consequence comes from any disparity in the numbers of Blues and Greens, creating a majority and a minority. In this case, while both groups suffer from the lack of mutual cooperation, the members of the minority group suffer more. No wonder minorities often seek defensive isolation.

To see why, suppose that there are eighty Greens and twenty Blues in a town, and everyone interacts with everyone else once a week. Then for the Greens, most of their interactions are within their own group and hence result in mutual cooperation. But for the Blues, most of their interactions are with the other group (the Greens), and hence result in punishing mutual defection. Thus, the average score of the minority Blues is less than the average score of the majority Greens. This effect will hold even when there is a tendency for each group to associate with its own kind. The effect still holds because if there are a certain number of times a minority Blue meets a majority Green, this will represent a larger share of the minority's total interactions than it does of the majority's total interactions (Rytina and Morgan 1982). The result is that labels can support stereotypes by which everyone suffers, and the minority suffers more than the rest.

Labels can lead to another effect as well. They can support status hierarchies. For example, suppose that everyone has some characteristic, such as height or strength or skin tone, that can be readily observed and that allows a comparison between two people. For simplicity imagine that there are no tie values, so that when two people meet it is clear which one has more of the characteristic and which one has less. Now suppose that everyone is a bully toward those beneath them and meek toward those above them. Can this be stable?

Yes, and here is an illustration. Suppose everyone uses the following strategy when meeting someone beneath them: alternate defection and cooperation unless the other player defects even once, in which case never cooperate again. This is being a bully in that you are often defecting, but never tolerating a defection from the other player. And suppose that everyone uses the following strategy when meeting someone above them: cooperate unless the other defects twice in a row, in which case never cooperate again. This is being meek in that you are tolerating being a sucker on alternating moves, but it is also being provocable in that you are not tolerating more than a certain amount of exploitation.

This pattern of behavior sets up a status hierarchy based on the observable characteristic. The people near the top do well because they can lord it over nearly everyone. Conversely, the people near the bottom are doing poorly because they are being meek to almost everyone. It is easy to see why someone near the top is happy with the social structure, but is there anything someone near the bottom can do about it acting alone?

Actually there isn't. The reason is that when the discount parameter is high enough, it would be better to take one's medicine every other move from the bully than to defect and face unending punishment.[2] Therefore, a person at the bottom of the social structure is trapped. He or she is doing poorly, but would do even worse by trying to buck the system.

The futility of isolated revolt is a consequence of the immutability of the other players' strategies. A revolt by a low-status player would actually hurt *both* sides. If the higher-status players might alter their behavior under duress, then this fact should be taken into account by a lower-status player contemplating revolt. But this consideration leads the higher-status players to be concerned with their reputation for firmness. To study this type of phenomena, one needs to look at the dynamics of reputations.

Reputation and Deterrence

A player's reputation is embodied in the beliefs of others about the strategy that player will use. A reputation is typically established through observing the actions of that player when interacting with other players. For example, Britain's reputation for being provocable was certainly enhanced by its decision to take back the Falkland Islands in response to the Argentine invasion. Other nations could observe Britain's decisions and make inferences about how it might react to their own actions in the future. Especially relevant would be Spanish inferences about the British commitment to Gibraltar, and Chinese inferences about the British commitment to Hong Kong. Whether these inferences would be correct is another matter. The point is that when third parties are watching, the stakes of the current situation expand from those immediately at hand to encompass the influence of the current choice on the reputations of the players.

Knowing people's reputations allows you to know something about what strategy they use even before you have to make your first choice. This possibility suggests the question of how valuable it would be to know for certain what strategy the other player is about to use with you. A way to measure the value of any piece of information is to calculate how much better you could do with the information than without it (Raiffa 1968). Thus, the better you can do without the information, the less you need the information, and the less it is worth. In both rounds of the Prisoner's Dilemma tournament, for example, TIT FOR TAT did well without knowing the strategy to be employed by the other player. Knowing the other's strategy would have allowed a player to do substantially better in only a few cases. For example, if the other player's strategy were known to be TIT FOR TWO TATS (which defects only if the other defected on both of the previous two moves), it would be possible to do better than TIT FOR TAT did by alternating defection with cooperation. But there are not many exploitable strategies in either round of the tournament, so knowing the other's strategy in advance would not actually help you do much better than the all-purpose strategy of TIT FOR TAT. In fact, the smallness of the gain from knowing the other's strategy is just another measure of the robustness of TIT FOR TAT.

The question about the value of information can also be turned around: what is the value (or cost) of having other players know *your* strategy? The answer, of course, depends on exactly what strategy you are using. If you are using an exploitable strategy, such as TIT FOR TWO TATS, the cost can be substantial. On the other hand, if you are using a strategy that is best met with complete cooperation, then you might be glad to have your strategy known to the other. For example, if you were using TIT FOR TAT, you would be happy to have the other player appreciate this fact and adapt to it, provided, of course, that the shadow of the future is large enough so that the best response is a nice strategy. In fact, as has been said, one of the advantages of TIT FOR TAT is that it is easy for it to be recognized in the course of a game even if the player using it has not yet established a reputation.

Having a firm reputation for using TIT FOR TAT is advantageous to a player, but it is not actually the best reputation to have. The best reputation to have is the reputation for being a bully. The best kind of bully to be is one who has a reputation for squeezing the most out of the other player while not tolerating any defections at all from the other. The way to squeeze the most out of the other is to defect so often that the other player just barely prefers cooperating all the time to defecting all the time. And the best way to encourage cooperation from the other is to be known as someone who will never cooperate again if the other defects even once.

Fortunately, it is not easy to establish a reputation as a bully. To become known as a bully you have to defect a lot, which means that you are likely to provoke the other player into retaliation. Until your reputation is well established, you are likely to have to get into a lot of very unrewarding contests of will. For example, if the other player defects even once, you will be torn between acting as tough as the reputation you want to establish requires and attempting to restore amicable relations in the current interaction.

What darkens the picture even more is that the other player may also be trying to establish a reputation, and for this reason may be unforgiving of the defections you use to try to establish your own reputation. When two players are each trying to establish their reputations for use against other players in future games, it is easy to see that their own interactions can spiral downward into a long series of mutual punishments.

Each side has an incentive to pretend not to be noticing what the other is trying to do. Both sides want to appear to be untrainable so that the other will stop trying to bully them.

The Prisoner's Dilemma tournament suggests that a good way for a player to appear untrainable is for the player to use the strategy of TIT FOR TAT. The utter simplicity of the strategy makes it easy to assert as a fixed pattern of behavior. And the ease of recognition makes it hard for the other player to maintain an ignorance of it. Using TIT FOR TAT is an effective way of holding still and letting the *other* player do the adaptation. It refuses to be bullied, but does not do any bullying of its own. If the other player does adapt to it, the result is mutual cooperation. In fact, deterrence is achieved through the establishment of a reputation.

One purpose of having a reputation is to enable you to achieve deterrence by means of a credible threat. You try to commit yourself to a response that you really would not want to make if the occasion actually arose. The United States deters the Soviets from taking West Berlin by threatening to start a major war in response to such a grab. To make such a threat credible, the United States seeks to establish a reputation as a country that actually does carry out such guarantees, despite the short-run cost.

Vietnam had just such a meaning to the American government when the decision to commit major combat forces was being made in 1965. The dominance of the desire to maintain a reputation was expressed in a secret memo to Secretary of Defense Robert McNamara from his Assistant Secretary for International Security Affairs, John McNaughton, defining U.S. aims in South Vietnam:

> *U.S. aims:*
> 70 percent—To avoid a humiliating U.S. defeat (to our reputation as a guarantor).
> 20 percent—To keep SVN (and adjacent) territory from Chinese hands.
> 10 percent—To permit the people of SVN to enjoy a better, freer way of life. (Quoted in Sheehan and Kenworthy 1971, p. 432)

Maintaining deterrence through achieving a reputation for toughness is important not only in international politics, but also in many domestic functions of the government. While this book is mainly about situations without central authority, the framework actually applies to many situations involving an authority. The reason is that even the most effective governments cannot take the compliance of its citizens for granted. Instead, a government has strategic interactions with the governed, and these interactions often take the form of an iterated Prisoner's Dilemma.

The Government and the Governed

A government must deter its citizens from breaking the law. For example, to collect taxes effectively, a government must maintain a reputation for prosecuting tax evaders. The government often spends far more investigating and prosecuting evaders than it acquires from the penalties levied against them. The government's goal, of course, is to maintain a reputation for catching and prosecuting evaders to deter anyone contemplating tax evasion in the future. And what is true for tax collection is also true for many forms of policing: the key to maintaining compliant behavior from the citizenry is that the government remains able and willing to devote resources far out of proportion to the stakes of the current issue in order to maintain its reputation for toughness.

In the case of a government and its citizens, the social structure has a single central actor and many peripheral ones. A comparable social structure exists with a monopolist trying to deter entry into its market. Still another example is an empire trying to deter revolt by its provinces. In each case, the problem is to prevent challenges by maintaining a reputation for firmness in dealing with them. To maintain this reputation might well require meeting a particular challenge with a toughness out of all proportion to the stakes involved in that particular issue.

Even the most powerful government cannot enforce any rule it chooses. To be effective, a government must elicit compliance from the majority of the governed. To do this requires setting and enforcing the rules so that it pays for most of the governed to obey most of the time. An example of this fundamental problem occurs in the regulation of industrial pollution.

As modeled by Scholz (1983), the government regulatory agency and a regulated company are in an iterated Prisoner's Dilemma with each other. The company's choices at any point are to comply voluntarily with the rules or to evade them. The agency's choices are to adopt an enforcement mode in dealing with that particular company which is either flexible or coercive.

If the agency enforces with flexibility and the firm complies with the rules, then both the agency and the firm benefit from mutual cooperation. The agency benefits from the company's compliance, and the company benefits from the agency's flexibility. Both sides avoid expensive enforcement and litigation procedures. Society also gains the benefits of full compliance at low cost to the economy. But if the firm evades and the agency uses coercive enforcement, both suffer the punishing costs of the resultant legalistic relationship. The firm also faces a temptation to evade if the agency is using a flexible enforcement policy which is unlikely to penalize evasion. And the agency faces a temptation to use the strict enforcement mode with a complying company in order to get the benefits of enforcing even unreasonably expensive rules.

The agency can adopt a strategy such as TIT FOR TAT which would give the company an incentive to comply voluntarily and thereby avoid the retaliation represented by the coercive enforcement policy. Under suitable conditions of the payoff and discount parameters, the relationship between the regulated and the regulator could be the socially beneficial one of repeated voluntary compliance and flexible enforcement.

The new feature introduced by Scholz's model of the

interaction between the government and the governed is the additional choice the government has concerning the toughness of the standards. To set a tough pollution standard, for example, would make the temptation to evade very great. On the other hand, to set a very lenient standard would mean more allowable pollution, thereby lessening the payoff from mutual cooperation which the agency would attain from voluntary compliance. The trick is to set the stringency of the standard high enough to get most of the social benefits of regulation, and not so high as to prevent the evolution of a stable pattern of voluntary compliance from almost all of the companies.

In addition to making and enforcing standards, governments often settle disputes between private parties. A good example is the case of a divorce in which the court awards child custody to one parent, and imposes a requirement of child support payments upon the other parent. Such settlements are notorious for the unreliability of the consequent support payments. For this reason, it has been proposed that the future interactions between the parents be given a reciprocal nature by allowing the custodial parent to withdraw visitation privileges if the other parent falls behind in the support payments (Mnookin and Kornhauser 1979). This proposal could amount to placing the parents in an iterated Prisoner's Dilemma, and leaving them to work out strategies based upon reciprocity. Hopefully, the result would benefit the child by promoting a stable pattern of cooperation between the parents based upon reciprocity that traded reliable support payments for regular visitation privileges.

Governments relate not only to their own citizens, but to other governments as well. In some contexts, each government can interact bilaterally with any other government. An example is the control of international trade in which a country can impose trade restrictions upon imports from another country, for instance as a retaliation against unfair trade practices. But an interesting characteristic of governments that has not yet been taken into account is that they are based upon specific territories. In a pure territorial system, each player has only a few neighbors, and interacts only with these neighbors. The dynamic properties of this type of social structure are the subject of the next section.

Territoriality

Nations, businesses, tribes, and birds are examples of players which often operate mainly within certain territories. They interact much more with their neighbors than with those who are far away. Hence their success depends in large part on how well they do in their interactions with their neighbors. But neighbors can serve another function as well. A neighbor can provide a role model. If the neighbor is doing well, the behavior of the neighbor can be imitated. In this way successful strategies can spread throughout a population, from neighbor to neighbor.

Territories can be thought of in two completely different ways. One way is in terms of geography and physical space. For example, the live-and-let-live system in trench warfare might have spread from part of the front line to adjacent parts. Another way of thinking about territories is in terms of an abstract space of characteristics. For example, a business might market a soft drink with a certain amount of sugar and a certain amount of caffeine. The "neighbors" of this soft drink are other drinks on the market with a little more or less sugar, or a little more or less caffeine. Similarly, a political candidate might take a position on a liberal/conservative dimension and a position on an internationalism/isolationism dimension. If there are many candidates vying with each other in an election, the "neighbors" of the candidate are those with similar positions. Thus territories can be abstract spaces as well as geographic spaces.

Colonization provides another mechanism in addition to imitation by which successful strategies can spread from place to place. Colonization would occur if the location of a less successful strategy was taken over by an offspring of a more successful neighbor. But whether strategies spread by imitation or colonization, the idea is the same: neighbors interact and the most successful strategy spreads to bordering locations. The individuals remain fixed in their locations, but their strategies can spread.

To make this process amenable to analysis, it must be formalized. For illustrative purposes, consider a simple structure of territories in which the entire territory is divided up so that each player has four neighbors, one to the north, one to the east, one to the south, and one to the west. In each "generation," each player attains a success score measured by its average performance with its four neighbors. Then if a player has one or more neighbors who are more successful, the player converts to the strategy of the most successful of them (or picks randomly among the best in case of a tie among the most successful neighbors).

Territorial social structures have many interesting properties. One of them is that it is at least as easy for a strategy to protect itself from a takeover by a new strategy in a territorial structure as it is in a nonterritorial structure. To see how this works, the definition of stability must be extended to include territorial systems. Recall from chapter 3 that a strategy can invade another if it can get a higher score than the population average in that environment. In other words, a single individual using a new strategy can invade a population of natives if the newcomer does better with a native than a native does with another native. If no strategy can invade the population of natives, then the native strategy is said to be collectively stable.[3]

To extend these concepts to territorial systems, suppose that a single individual using a new strategy is introduced into one of the neighborhoods of a population where everyone else is using a native strategy. One can say that the new strategy *territorially invades* the native strategy if every location in the territory will eventually convert to the new strategy. Then one can say that native strategy is *territorially stable* if no strategy can territorially invade it.

All this leads to a rather strong result: it is no harder for a strategy to be territorially stable than it is to be collectively stable. In other words, the conditions that are needed for a strategy to protect itself from takeover by an invader are no more stringent in a territorial social system than they are in a social system where anyone is equally likely to meet anyone else.

Proposition 8. If a rule is collectively stable, it is territorially stable.

The proof of this proposition gives some insight into the dynamics of territorial systems. Suppose there is a territorial system in which everyone is using a native strategy that is collectively stable, except for one individual who is using a new strategy. The situation is illustrated in figure 3. Now consider whether a neighbor of the newcomer would ever have reason to convert to the newcomer's strategy. Since

FIGURE 3
A Portion of a Territorial Social Structure with a Single Mutant

B	B	B	B	B
B	B	B	B	B
B	B	A	B	B
B	B	B	B	B
B	B	B	B	B

the native strategy is collectively stable, the newcomer cannot be scoring as well when surrounded by natives as a native who is surrounded by natives is scoring. But every neighbor of the newcomer actually does have a neighbor who is also a native and who is entirely surrounded by other natives. Therefore no neighbor of the newcomer will find the newcomer to be the most successful neighbor to imitate. So all of the newcomer's neighbors will retain their own native strategy, or, what amounts to the same thing, will convert to the strategy of one of their native neighbors. Therefore, the new strategy cannot spread in a population of collectively stable strategies, and consequently a collectively stable strategy is also territorially stable.

The proposition that a collectively stable rule is territorially stable demonstrates that protection from invasion is at least as easy in a territorial system as in a freely mixing system. One implication is that mutual cooperation can be sustained in a territorial system by a nice rule with no greater requirement on the size of the discount parameter relative to the payoff parameters than it takes to make that nice rule collectively stable.

Even with the help of a territorial social structure to maintain stability, a nice rule is not necessarily safe. If the shadow of the future is sufficiently weak, then no nice

FIGURE 4
Meanies Spreading in a Population of TIT FOR TAT

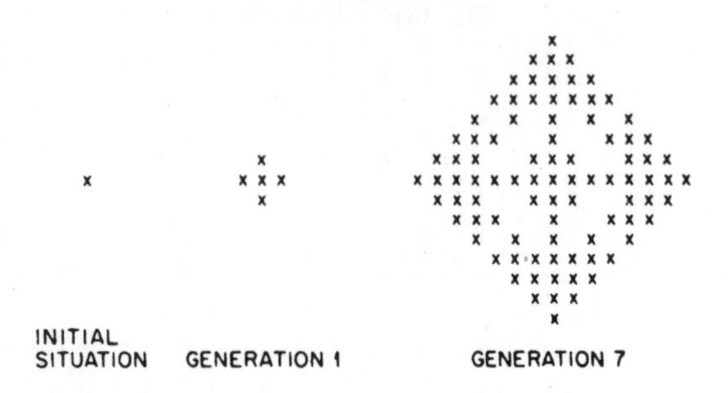

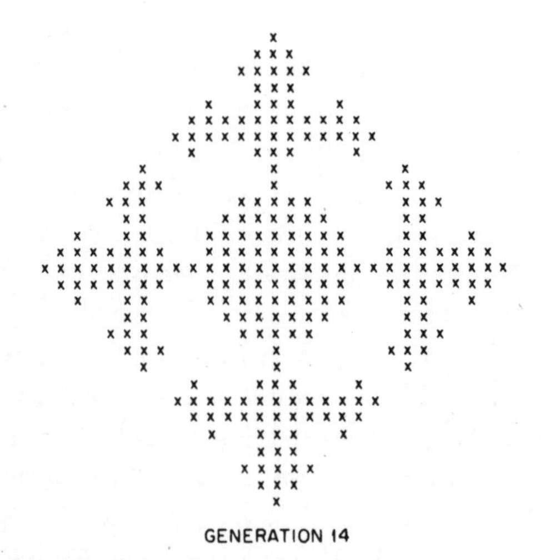

strategy can resist invasion even with the help of territoriality. In such a case, the dynamics of the invasion process can sometimes be extremely intricate and quite fascinating to look at. Figure 4 shows an example of such an intricate pattern. It represents the situation of a single player who always defects invading a territorial population of individuals using TIT FOR TAT. In this case, the shadow of

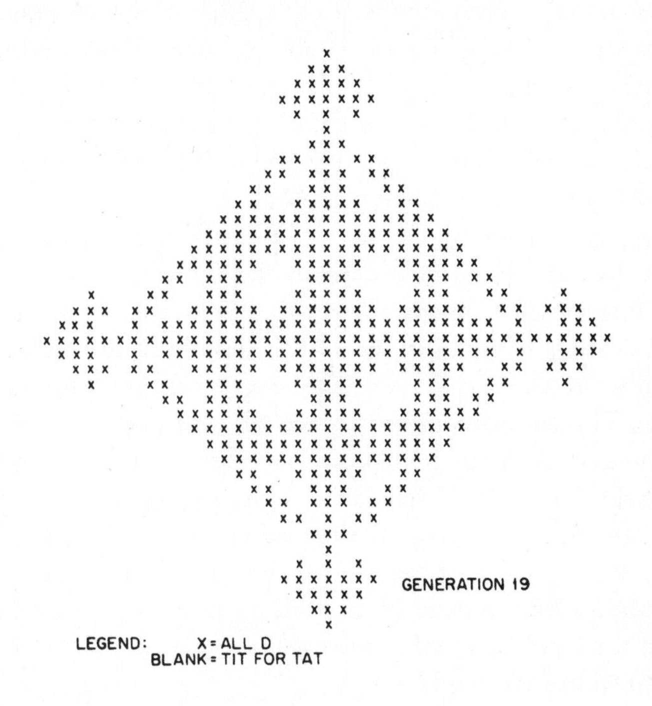

the future has been made quite weak, as reflected in low value of the discount parameter, $w = 1/3$. The four payoff parameters have been selected to provide an illustration of the intricacies that are possible. In this case $T = 56$, $R = 29$, $P = 6$, and $S = 0$.[4] With these values, figure 4 shows what happens after one, seven, fourteen, and nineteen generations. The meanies colonize the original TIT FOR TAT population, forming a fascinating pattern of long borders and bypassed islands of cooperators.

Another way of looking at the effects of territoriality is to investigate what happens when the players are using a wide variety of more or less sophisticated strategies. A convenient way to do this is to use the sixty-three different rules available from the second round of the Computer Tournament. Assigning each rule to four territories allows exactly the right number of players to fill a space which is 14 cells high and 18 cells wide. To guarantee that everyone still has exactly four neighbors, the borders of the space can be thought of as wrapping around upon themselves. For example, a square on the far right has as one of its neighbors the corresponding square on the far left.

To see what happens when the players are using such a wide variety of more or less sophisticated decision rules, it is only necessary to simulate the process one generation at a time. The tournament results provide the necessary information about the score that each rule gets with any particular neighbor it might have. The score of a territory is then the average of its scores with the four rules that neighbor it. Once the score of each territory is established, the conversion process begins. Each territory that has a more successful neighbor simply converts to the rule of the most successful of its neighbors.

To be sure that the results were not too sensitive to the particular random assignments that began the process, the whole simulation was repeated ten times with different random assignments each time. Each simulation was conducted generation after generation until there were no further conversions. This took from eleven to twenty-four generations. In each case, the process stopped evolving only when all of the rules that were not nice had been eliminated. With only nice rules left, everyone was always cooperating with everyone else and no further conversions would take place.

A typical final pattern is shown in figure 5. There are a number of striking features in this stable pattern of strate-

FIGURE 5

Example of a Final Population in a Territorial System

6	6	6	1	44	44	44	44	44	6	6	7	7	6	7	6	6	6
6	1	31	1	1	44	44	44	44	3	6	3	6	6	6	6	6	6
6	6	31	31	1	1	1	1	1	1	3	3	3	52	52	6	6	6
6	1	31	31	31	31	31	31	31	31	31	3	3	6	6	6	6	6
6	9	31	31	31	31	31	31	31	31	31	31	3	6	6	6	6	6
6	31	31	31	31	31	31	31	31	31	31	31	31	6	6	6	6	6
6	31	31	6	6	9	31	31	6	9	41	31	31	6	6	31	31	6
6	31	31	6	6	9	9	9	6	41	41	31	4	31	31	31	31	6
6	31	31	9	9	9	9	9	6	41	41	41	41	31	31	31	31	31
6	31	6	9	9	9	6	6	6	41	41	41	17	31	31	31	31	31
6	6	9	7	9	9	6	6	6	41	41	41	41	31	31	31	31	7
6	6	7	7	7	6	6	6	9	41	41	7	7	7	7	7	7	6
6	6	7	7	7	6	6	6	6	41	6	7	7	7	7	7	7	6
6	6	7	7	6	6	44	6	6	6	6	7	7	7	7	6	6	6

Legend: The numbers at each location indicate the rank order of the strategy in round two of the Computer Prisoner's Dilemma Tournament. For example 1=TIT FOR TAT, and 31=NYDEGGER.

gies. In the first place, the surviving strategies are generally clumped together into regions of varying size. The random scattering that began the population has largely given way to regions of identical rules which sometimes spread over a substantial distance. Yet there are also a few very small regions and even single territories surrounded by two or three different regions.

The rules which survived tend to be rules which scored well in the tournament. For example, TIT FOR TAT was represented an average of seventeen times in the final population, after having started with exactly four copies in each run. But there were also five other rules which had better representation in the final populations. The best of these was a ruled submitted by Rudy Nydegger which ranked only thirty-first among the sixty-three rules in the round robin tournament. In the territorial system it finished with an average of forty followers. Thus, a rule that wound up right in the middle of the round robin tournament standings was by far the most successful rule in the two-dimensional territorial system. How could this have happened?

The strategy of the rule itself is hard to analyze because it is based upon a complex table-lookup scheme which uses the previous three outcomes to decide what to do next. But the performance of the rule can be analyzed in terms of how it actually fared with each of the rules it could meet. Like the other rules which survived, NYDEGGER never defects first. But what is unique about it is that when the other player defects first, NYDEGGER is sometimes able to get the other player to "apologize" so profusely that NYDEGGER actually ends up with a higher score than if there had just been mutual cooperation. This happens with five of the twenty-four rules which are not nice. In the round robin tournament, this is not enough to do very well since NYDEGGER often gets in trouble with the other rules which are not nice.

In the territorial system, things work differently. By getting five of the rules which are not nice to apologize, NYDEGGER wins a great many converts from its neighbors. When one of these apologizers is next to NYDEGGER and the other three neighbors are nice rules, NYDEGGER is likely to do better than any of its four neighbors or even any of *their* neighbors. In this way, it can convert not only the apologist, but some or all of its other neighbors as well. Thus, in a social system based on diffusion by imitation, there is a great advantage to being able to attain outstanding success, even if it means that the average rate of success is not outstanding. This is because the occasions of outstanding success win many converts. The fact that NY-

DEGGER is nice means that it avoids unnecessary conflict, and continues to hold its own when the rules which are not nice are eliminated. The advantage that NYDEGGER gets is based on the fact that while five rules are abjectly apologetic to it, no other nice rule elicits such apologies from more than two other rules.

The territorial system demonstrates quite vividly that the way the players interact with each other can affect the course of the evolutionary process. A variety of structures have now been analyzed in evolutionary terms, although many other interesting possibilities await analysis.[5] Each of the five structures examined in this book reveal different facets of the evolution of cooperation:

1. Random mixing was used as the fundamental type of structure. The round robin tournaments and the theoretical propositions showed how cooperation based upon reciprocity can thrive even in a setting with such a minimal social structure.

2. Clusters of players were examined to see how the evolution of cooperation could have gotten started in the first place. Clusters allow a newcomer to have at least a small chance of meeting another newcomer, even though the newcomers themselves are a negligible part of the whole environment of the natives. Even if most of a newcomer's interactions are with uncooperative natives, a small cluster of newcomers who use reciprocity can invade a population of meanies.

3. Differentiation of the population was shown to occur when the players have more information about each other than is contained in the history of their own interaction. If the players have labels indicating their group membership or personal attributes, stereotyping and status hierarchies can develop. If the players can observe each other interact-

ing with others, they can develop reputations; and the existence of reputations can lead to a world characterized by efforts to deter bullies.

4. Governments were found to have their own strategic problems in terms of achieving compliance from most of their citizens. Not only is this a problem of choosing an effective strategy to use in a particular case, but it is also a question of how to set the standards so that compliance will be both attractive to the citizen and beneficial to the society.

5. Territorial systems were examined to see what would happen if the players interacted only with their neighbors, and imitated a neighbor who was more successful than they were. Interactions with neighbors were found to give rise to intricate patterns in the spread of particular strategies, and to promote the growth of those strategies that scored unusually well in some settings even though they did poorly in others.

CHAPTER 9

The Robustness of Reciprocity

THE EVOLUTIONARY APPROACH is based on a simple principle: whatever is successful is likely to appear more often in the future. The mechanism can vary. In classical Darwinian evolution, the mechanism is natural selection based upon differential survival and reproduction. In Congress, the mechanism can be an increased chance of reelection for those members who are effective in delivering legislation and services for their constituency. In the business world, the mechanism can be the avoidance of bankruptcy by a profitable company. But the evolutionary mechanism need not be a question of life and death. With intelligent players, a successful strategy can appear more often in the future because other players convert to it. The conversion can be based on more or less blind imitation of the successful players, or it can be based on a more or less informed process of learning.

The evolutionary process needs more than differential growth of the successful. In order to go very far it also needs a source of variety—of new things being tried. In the genetics of biology, this variety is provided by mutation and by a reshuffling of genes with each generation. In social processes, the variety can be introduced by the "trial" in "trial and error" learning. This kind of learning might or might not reflect a high degree of intelligence. A new pattern of behavior might be undertaken simply as a random variant of an old pattern of behavior, or the new strategy could be deliberately constructed on the basis of prior experience and a theory about what is likely to work best in the future.

To study different aspects of the evolutionary process, different methodological tools have been used. One set of questions asked about the destination of the evolutionary process. To study this, the concept of collective (or evolutionary) stability was used to study where the evolutionary process would stop. The idea was to determine which strategies could not be invaded if they were used by everyone. The virtue of this approach is that it allowed a good specification of which types of strategies can protect themselves, and under what conditions this protection can work. For example, it was shown that TIT FOR TAT would be collectively stable if the shadow of the future were large enough, and that the strategy of always defecting would be collectively stable under all possible conditions.

The power of the collective stability approach is that it allows a consideration of *all* possible new strategies, whether minor variants of the common strategy or completely new ideas. The limitation of the stability approach is that it only tells what will last once established, but it does not tell what will get established in the first place. Since many different strategies can be collectively stable once established in a population, it is important to know which strategies are likely to get established in the first place. For this a different methodology was needed.

To see what is likely to get established in the first place, the emphasis must be placed upon the variety of things that can happen at once in a population. To capture this variety, the tournament approach was used. The tournament itself was conducted to encourage the presence of sophisticated strategies, which were attained in the first round by soliciting entries from professional game theorists. Refinement of strategies was carried further in the second round by making sure that the new entrants were aware of the results of the first round. Thus, new ideas could enter the tournament either as refinements of the old ideas or as totally new conceptions of what might work best. Then the analysis of what actually worked best in this variegated environment told a great deal about which kind of strategy is likely to flourish.

Since the process of getting fully established is likely to take a considerable amount of time, another kind of technique was used to study the changing prospects of strategies as their social environment changes. This technique was an ecological analysis, which calculated what would happen if each generation had strategies growing in frequency in proportion to their success in the previous generation. This was an ecological approach because it introduced no new strategies, but instead determined the consequences over hundreds of generations of the variety of strategies already represented in the tournament. It allowed for an analysis of whether the strategies that were successful in the beginning would remain successful after the poor performers had dropped out. The growth of the successful strategies in each generation could be thought of as due to either better survival and reproduction of the users of that strategy, or due to a greater chance of being imitated by the others.

Related to the ecological analysis was the territorial analysis of what would happen if the sixty-three strategies of the second round of the tournament were scattered in a territorial structure, with the player at each location interacting with the four neighbors of that location. In the territorial system, determination of what is successful is local. Each location which has a more successful neighbor adopts the strategy of the most successful of its neighbors. As in the ecological simulation, this growth of the more successful can be attributed to either better survival and reproduction, or to a greater chance of being imitated by others.

To use these tools of evolutionary analysis, what is needed is a way to determine how any given strategy will perform with any other given strategy. In simple cases, this calculation can be done algebraically, as in the determination of how TIT FOR TAT will do when it meets a player who always defects. In more complex cases, the calculation can be done by simulating the interactions and cumulating the payoffs received, as in the conduct of the Computer Tournament for the Prisoner's Dilemma. The ideas of a time discount and uncertain ending of the interaction were incorporated in the tournament by varying the lengths of the games. The consequences of the probabilistic nature of some strategies were handled by averaging over several repetitions of the interaction between the same two strategies.

These tools of evolutionary analysis could be used with any social setting. In this book they have been applied to one particular kind of social setting, a setting which captures the fundamental dilemma of cooperation. The potential for cooperation arises when each player can help the other. The dilemma arises when giving this help is costly. The opportunity for mutual gain from cooperation comes into play when the gains from the other's cooperation are larger than the costs of one's own cooperation. In that case mutual cooperation is preferred by both to mutual noncooperation (so-called defection). But getting what you prefer is not so easy. There are two reasons. In the first place, you have to get the other player to help—even though the other player is better off in the short run by not helping. In the second place, you are tempted to get whatever help you can without providing any costly help yourself.[1]

The main results of Cooperation Theory are encouraging. They show that cooperation can get started by even a small cluster of individuals who are prepared to reciprocate cooperation, even in a world where no one else will cooperate. The analysis also shows that the two key requisites for cooperation to thrive are that the cooperation be based on reciprocity, and that the shadow of the future is important enough to make this reciprocity stable. But once cooperation based on reciprocity is established in a population, it can protect itself from invasion by uncooperative strategies.

It is encouraging to see that cooperation can get started, can thrive in a variegated environment, and can protect itself once established. But what is most interesting is how little had to be assumed about the individuals or the social setting to establish these results. The individuals do not have to be rational: the evolutionary process allows the successful strategies to thrive, even if the players do not know why or how. Nor do the players have to exchange messages or commitments: they do not need words, because their deeds speak for them. Likewise, there is no need to assume trust between the players: the use of reciprocity can be enough to make defection unproductive. Altruism is not needed: successful strategies can elicit cooperation even from an egoist. Finally, no central authority is needed: cooperation based on reciprocity can be self-policing.

The emergence, growth, and maintenance of cooperation do require some assumptions about the individuals and the social setting. They require an individual to be able to recognize another player who has been dealt with before. They also require that one's prior history of interactions with this player can be remembered, so that a player can be responsive. Actually, these requirements for recognition and recall are not as strong as they might seem. Even bacteria can fulfill them by interacting with only one other organism and using a strategy (such as TIT FOR TAT) which responds only to the recent behavior of the other player. And if bacteria can play games, so can people and nations.

For cooperation to prove stable, the future must have a sufficiently large shadow. This means that the importance of the next encounter between the same two individuals must be great enough to make defection an unprofitable strategy when the other player is provocable. It requires that the players have a large enough chance of meeting again and that they do not discount the significance of their next meeting too greatly. For example, what made cooperation possible in the trench warfare of World War I was the fact that the same small units from opposite sides of no-man's-land would be in contact for long periods of time, so that if one side broke the tacit understandings, then the other side could retaliate against the same unit.

Finally, the evolution of cooperation requires that successful strategies can thrive and that there be a source of variation in the strategies which are being used. These mechanisms can be classical Darwinian survival of the fittest and the mutation, but they can also involve more deliberate processes such as imitation of successful patterns of behavior and intelligently designed new strategic ideas.

In order for cooperation to get started in the first place, one more condition is required.The problem is that in a world of unconditional defection, a single individual who offers cooperation cannot prosper unless others are around who will reciprocate. On the other hand, cooperation can emerge from small clusters of discriminating individuals as long as these individuals have even a small proportion of their interactions with each other. So there must be some clustering of individuals who use strategies with two properties: the strategies will be the first to cooperate, and they will discriminate between those who respond to the cooperation and those who do not.

The conditions for the evolution of cooperation tell what is necessary, but do not, by themselves, tell what strategies will be most successful. For this question, the tournament approach has offered striking evidence in favor of the robust success of the simplest of all discriminating strategies: TIT FOR TAT. By cooperating on the first move, and then doing whatever the other player did on the previous move, TIT FOR TAT managed to do well with a wide variety of more or less sophisticated decision rules. It not only won the first round of the Computer Prisoner's Dilemma Tournament when facing entries submitted by professional game theorists, but it also won the second round which included over sixty entries designed by people who were able to take the results of the first round into account. It was also the winner in five of the six major variants of the second round (and second in the sixth variant). And most impressive, its success was not based only upon its ability to do well with strategies which scored poorly for themselves. This was shown by an ecological analysis of hypothetical future rounds of the tournament. In this simulation of hundreds of rounds of the tournament, TIT FOR TAT again was the most successful rule, indicating that it can do well with good and bad rules alike.

TIT FOR TAT's robust success is due to being nice, provocable, forgiving, and clear. Its niceness means that it is never the first to defect, and this property prevents it from getting into unnecessary trouble. Its retaliation discourages the other side from persisting whenever defection is tried. Its forgiveness helps restore mutual cooperation. And its clarity makes its behavioral pattern easy to recognize; and once recognized, it is easy to perceive that the best way of dealing with TIT FOR TAT is to cooperate with it.

Despite its robust success, TIT FOR TAT cannot be called the ideal strategy to play in the iterated Prisoner's Dilemma. For one thing, TIT FOR TAT and other nice rules require for their effectiveness that the shadow of the future be sufficiently great. But even then there is no ideal strategy independent of the strategies used by the others. In some extreme environments, even TIT FOR TAT would do poorly—as would be the case if there were not enough others who would ever reciprocate its initial cooperative choice. And TIT FOR TAT does have its strategic weaknesses as well. For example, if the other player defects once, TIT FOR TAT will always respond with a defection, and then if the other player does the same in response, the result would be an unending echo of alternating defections. In this sense, TIT FOR TAT is not forgiving enough. But another problem is that TIT FOR TAT is too forgiving to those rules which are totally unresponsive, such as a completely random rule. What can be said for TIT FOR TAT is that it does indeed perform well in a wide variety of settings where the other players are all using more or less sophisticated strategies which themselves are designed to do well.

If a nice strategy, such as TIT FOR TAT, does eventually come to be adopted by virtually everyone, then individuals using this nice strategy can afford to be generous in dealing with any others. In fact, a population of nice rules can also protect itself from clusters of individuals using any other strategy just as well as they can protect themselves against single individuals.

These results give a chronological picture for the evolution of cooperation. Cooperation can begin with small clusters. It can thrive with rules that are nice, provocable, and somewhat forgiving. And once established in a population, individuals using such discriminating strategies can protect themselves from invasion. The overall level of cooperation tends to go up and not down. In other words, the machinery for the evolution of cooperation contains a ratchet.

The operation of this ratchet was seen in the development of the norm of reciprocity in the United States Congress. As described in the first chapter, in the early days of the republic, members of Congress were known for their deceit and treachery. They were quite unscrupulous and frequently lied to each other. Yet, over the years, cooperative patterns of behavior emerged and proved stable. These patterns were based upon the norm of reciprocity.

Many other institutions have developed stable patterns of cooperation based upon similar norms. Diamond markets, for example, are famous for the way their members exchange millions of dollars worth of goods with only a verbal pledge and a handshake. The key factor is that the participants know they will be dealing with each other again and again. Therefore any attempt to exploit the situation will simply not pay.

A wonderful illustration of this principle is provided in the memoirs of Ron Luciano, a baseball umpire who sometimes had his "bad days."

> Over a period of time I learned to trust certain catchers so much that I actually let them umpire for me on the bad days. The bad days usually followed the good nights. . . . On those days there wasn't much I could do but take two aspirins and call as little as possible. If someone I trusted was catching . . . I'd tell them, "Look, it's a bad day. You'd better take it for me. If it's a strike, hold your glove in place for an extra second. If it's a ball, throw it right back. And please, don't yell."

This reliance on the catcher could work because if Luciano ever suspected that he was being taken advantage of, he would have many opportunities to retaliate.

> No one I worked with ever took advantage of the situation, and no hitter ever figured out what I was doing. And only once, when Ed Herrman was calling the pitches, did a pitcher ever complain about a call. I smiled; I laughed; but I didn't say a word. I was tempted, though, I was really tempted. (Luciano and Fisher 1982, p. 166)

Ordinary business transactions are also based upon the idea that a continuing relationship allows cooperation to develop without the assistance of a central authority. Even though the courts do provide a central authority for the resolution of business disputes, this authority is usually not invoked. A common business attitude is expressed by a purchasing agent who said that "if something comes up you get the other man on the telephone and deal with the problem. You don't read legalistic contract clauses at each other if you ever want to do business again" (Macaulay 1963, p. 61). This attitude is so well established that when a large manufacturer of packaging materials inspected its records it found that it had failed to create legally binding contracts in two-thirds of the orders from its customers (Macaulay 1963). The fairness of the transactions is guaranteed not by the threat of a legal suit, but rather by the anticipation of mutually rewarding transactions in the future.

It is precisely when this anticipation of future interaction breaks down that an external authority is invoked. According to Macaulay, perhaps the most common type of business contracts case fought all the way to the appellate courts is an action for a wrongful termination of a dealer's franchise by a parent company. This pattern of conflict makes sense because once a franchise is ended, there is no prospect for further mutually rewarding transactions between the franchiser and the parent company. Cooperation ends, and costly court battles are often the result.

In other contexts, mutually rewarding relations become so commonplace that the separate identities of the participants can become blurred. For example, Lloyd's of London began as a small group of independent insurance brokers. Since the insurance of a ship and its cargo would be a large undertaking for one dealer, several brokers frequently made trades with each other to pool their risks. The frequency of the interactions was so great that the underwriters gradually developed into a federated organization with a formal structure of its own.

The importance of future interactions can provide a guide to the design of institutions. To help promote cooperation among members of an organization, relationships should be structured so that there are frequent and durable interactions among specific individuals. Corporations and bureaucracies are often structured in just this way, as discussed in chapter 8.

Sometimes the problem is one of retarding rather than promoting cooperation. An example is the prevention of collusive business practices by avoiding the very conditions which would promote cooperation. Unfortunately, the very ease with which cooperation can evolve even among egoists suggests that the prevention of collusion is not an easy task. Cooperation certainly does not require formal agreements or even face-to-face negotiations. The fact that cooperation based upon reciprocity can emerge and prove stable suggests that antitrust activities should pay more attention to preventing the conditions that foster collusion than to searching for secret meetings among executives of competing firms.

Consider, for example, the practice of the government selecting two companies for competitive development contracts for a new military airplane. Since aerospace companies specialize to some degree in planes for either the Air Force or the Navy, there is a tendency for firms with the same specialty to face each other in the final competition (Art 1968). This frequency of interaction between two given companies makes tacit collusion relatively easy to achieve. To make tacit collusion more difficult, the government should seek methods of reducing specialization or compensating for its effects. Pairs of companies which shared a specialization would then expect to interact less often in the final competitions. This would cause later interactions between them to be worth relatively less, reducing the shadow of the future. If the next expected interaction is sufficiently far off, reciprocal cooperation in the form of tacit collusion ceases to be a stable policy.

The potential for attaining cooperation without formal agreements has its bright side in other contexts. For example, it means that cooperation on the control of the arms race does not have to be sought entirely through the formal mechanism of negotiated treaties. Arms control could also evolve tacitly. Certainly, the fact that the United States and the Soviet Union know that they will both be dealing with each other for a very long time should help establish the necessary conditions. The leaders may not like each other, but neither did the soldiers in World War I who learned to live and let live.

Occasionally a political leader gets the idea that cooperation with another major power should not be sought because a better plan would be to drive them into bankruptcy. This is an extraordinarily risky enterprise because the target need not limit its response to the withholding of normal cooperation, but would also have a strong incentive to escalate the conflict before it was irreversibly weakened. Japan's desperate gamble at Pearl Harbor, for example, was a response to powerful American economic sanctions aimed at stopping Japanese intervention in China (Ike 1967; Hosoya 1968). Rather than give up what it regarded as a vital sphere, Japan decided to attack America before becoming even further weakened. Japan understood that America was much more powerful, but decided that the cumulative effects of the sanctions made it better to attack rather than to wait for the situation to get even more desperate.

Trying to drive someone bankrupt changes the time perspective of the participants by placing the future of the interaction very much in doubt. And without the shadow of the future, cooperation becomes impossible to sustain. Thus, the role of time perspectives is critical in the maintenance of cooperation. When the interaction is likely to continue for a long time, and the players care enough about their future together, the conditions are ripe for the emergence and maintenance of cooperation.

The foundation of cooperation is not really trust, but the durability of the relationship. When the conditions are right, the players can come to cooperate with each other through trial-and-error learning about possibilities for mutual rewards, through imitation of other successful players, or even through a blind process of selection of the more successful strategies with a weeding out of the less successful ones. Whether the players trust each other or not is less important in the long run than whether the conditions are ripe for them to build a stable pattern of cooperation with each other.

Just as the future is important for the establishment of the conditions for cooperation, the past is important for the monitoring of actual behavior. It is essential that the players are able to observe and respond to each other's prior choices. Without this ability to use the past, defections could not be punished, and the incentive to cooperate would disappear.

Fortunately, the ability to monitor the prior behavior of the other player does not have to be perfect. The Computer Tournament for the Prisoner's Dilemma assumed perfect knowledge of the other player's prior choices. In many settings, however, a player may occasionally misperceive the choice made by the other. A defection may go undetected, or a cooperation may be misinterpreted as a defection. To explore the implications of misperception, the first round of the tournament was run again with the modification that every choice had a 1 percent chance of being misperceived by the other player. As expected, these misunderstandings resulted in a good deal more defection between the players. A surprise was that TIT FOR TAT was still the best decision rule. Although it got into a lot of trouble when a single misunderstanding led to a long echo of alternating retaliations, it could often end the echo with another misperception. Many other rules were less forgiving, so that once they got into trouble, they less often got out of it. TIT FOR TAT did well in the face of misperception of the past because it could readily forgive and thereby have a chance to reestablish mutual cooperation.

The role of time perspective has important implications for the design of institutions. In large organizations, such as business corporations and governmental bureaucracies, executives are often transferred from one position to another approximately every two years.[2] This gives executives a strong incentive to do well in the short run, regardless of the consequences for the organization in the long run. They know that soon they will be in some other position, and the consequences of their choices in their previous post are not likely to be attributed to them after they have left their position. This gives two executives a mutual incentive to defect when either of their terms is drawing to an end. The result of rapid turnover could therefore be a lessening of cooperation within the organization.

As pointed out in chapter 3, a similar problem arises when a political leader appears to have little chance of reelection. The problem becomes even more acute with a lame duck. From the point of view of the public, a politician facing an end of career can be dangerous because of the increased temptation to seek private goals rather than maintain a pattern of cooperation with the electorate for the attainment of mutually rewarding goals.

Since the turnover of political leaders is a necessary part of democratic control, the problem must be solved another way. Here, political parties are useful because they can be held accountable by the public for the acts of their elected members. The voters and the parties are in a long-term relationship, and this gives the parties an incentive to select candidates who will not abuse their responsibilities. And if a leader is discovered giving in to temptation, the voters can take this into account in evaluating the other candidates of the same party in the next election. The punishment of the Republican party by the electorate after Watergate shows that parties are indeed held responsible for the defections of their leaders.

In general, the institutional solutions to turnover need to involve accountability beyond the individual's term in a particular position. In an organizational or business setting, the best way to secure this accountability would be to keep track not only of the person's success in that position, but also the state in which the position was left to the next occupant. For example, if an executive sought a quick gain by double-crossing a colleague just before transferring to a new plant, this fact should be taken into account in evaluating that executive's performance.

Cooperation Theory has implications for individual choice as well as for the design of institutions. Speaking personally, one of my biggest surprises in working on this project has been the value of provocability. I came to this project believing one should be slow to anger. The results of the Computer Tournament for the Prisoner's Dilemma demonstrate that it is actually better to respond quickly to a provocation. It turns out that if one waits to respond to uncalled for defections, there is a risk of sending the wrong signal. The longer defections are allowed to go unchallenged, the more likely it is that the other player will draw the conclusion that defection can pay. And the more strongly this pattern is established, the harder it will be to break it. The implication is that it is better to be provocable sooner, rather than later. The success of TIT FOR TAT certainly illustrates this point. By responding right away, it gives the quickest possible feedback that a defection will not pay.

The response to potential violations of arms control agreements illustrates this point. The Soviet Union has occasionally taken steps which appear to be designed to probe the limits of its agreements with the United States. The sooner the United States detects and responds to these Soviet probes, the better. Waiting for them to accumulate only risks the need for a response so large as to evoke yet more trouble.

The speed of response depends upon the time required to detect a given choice by the other player. The shorter this time is, the more stable cooperation can be. A rapid detection means that the next move in the interaction comes quickly, thereby increasing the shadow of the future as represented by the parameter w. For this reason the only arms control agreements which can be stable are those whose violations can be detected soon enough. The critical requirement is that violations can be detected before they can accumulate to such an extent that the victim's provocability is no longer enough to prevent the challenger from having an incentive to defect.

The tournament results concerning the value of provocability are complemented by the theoretical analysis of what it takes for a nice rule to be collectively stable. In order for a nice rule to be able to resist invasion, the rule must be provocable by the very first defection of the other player (proposition 4 in chapter 3). Theoretically, the response need not come immediately, and it need not occur with certainty, but it must have a real probability of coming eventually. The important thing is that the other player does not wind up having an incentive to defect.

Of course, provocability has a danger. The danger is that if the other player does try a defection, retaliation will lead to further retaliation, and the conflict will degenerate into an unending string of mutual defections. This can certainly be a serious problem. For example, in many cultures blood feuds between clans can continue undiminished for years and even generations (Black-Michaud 1975).

This continuation of the conflict is due to the echo effect: each side responds to the other's last defection with a new defection of its own. One solution is to find a central authority to police both sides, imposing a rule of law. Unfortunately this solution is often not available. And even when there is a rule of law, the costs of using the courts for routine affairs such as enforcement of business contracts can be prohibitive. When the use of a central authority is impossible or too expensive, the best method is to rely on a strategy which will be self-policing.

Such a self-policing strategy must be provocable, but the response must not be too great lest it lead to an unending echo of defections. For example, suppose that the Soviet Union in conjunction with the other Warsaw Pact countries undertakes a partial mobilization of its armed forces throughout Eastern Europe. This mobilization would give the Soviets an added advantage if conventional war were to break out. A useful response from NATO would be to increase its own state of alert. If additional troops moved from the Soviet Union to Eastern Europe, NATO should respond with additional troops moved from the United States. Betts (1982, pp. 293–94) recommends that this type of response be automatic so that it can be made clear to the Soviets that such increases in NATO readiness are standard procedure and take place only after Soviet mobilization. He also recommends that the response be limited, say one American division moved for every three Soviet divisions mobilized. In effect, this would help limit the echo effects.

Limited provocability is a useful feature of a strategy designed to achieve stable cooperation. While TIT FOR TAT responds with an amount of defection exactly equal to the other's defection, in many circumstances the stability of cooperation would be enhanced if the response were slightly less than the provocation. Otherwise, it would be all too easy to get into a rut of unending responses to each other's last defection. There are several ways for an echo effect to be controlled. One way is for the player who first defected to realize that the other's response need not call for yet another defection. For example, the Soviets might realize that NATO's mobilization was merely a response to their own, and hence need not be regarded as threatening. Of course the Soviets might not see it that way, even if the NATO response was automatic and predictable. Therefore, it is also useful if the NATO response is somewhat less than proportional to the Soviet mobilization. Then if the Soviet response is also somewhat less than the NATO mobilization, the escalation of preparations can become stabilized, and then possibly reversed for a return to normal.

Fortunately, friendship is not necessary for cooperation to evolve. As the trench warfare example demonstrates, even antagonists can learn to develop cooperation based upon reciprocity. The requirement for the relationship is not friendship, but durability. The good thing about international relations is that the major powers can be quite certain they will be interacting with each other year after year. Their relationship may not always be mutually rewarding, but it *is* durable. Therefore, next year's interactions should cast a large shadow on this year's choices, and cooperation has a good chance to evolve eventually.

Foresight is not necessary either, as the biological examples demonstrate. But without foresight, the evolutionary process can take a very long time. Fortunately, humans do have foresight and use it to speed up what would otherwise be a blind process of evolution. The most striking example of this was the difference between the first and second rounds of the Computer Prisoner's Dilemma Tournament. In the first round the contestants were professional game theorists who represented the state of the art in the understanding of how to do well in the iterated Prisoner's Dilemma. When their rules were paired with each other, the result was an average score per move of 2.10 which is only slightly better than halfway from $P=1$ (the punishment for mutual defection) to $R=3$ (the reward for mutual cooperation). The players in the second round did much better, scoring 2.60, which is a little better than three-quarters of the way from the mutual punishment to the mutual reward.[3] Thus, the players were able to use the results of the first round, including the value of reciprocity, to anticipate what would work well in the second round. On the whole, their foresight paid off with substantially higher scores.

The result was that the second round was more sophisticated than the first. Cooperation based upon reciprocity was firmly established. The various attempts at exploitation of the unsophisticated entries of the first round all failed in the environment of the second round, demonstrating that the reciprocity of strategies like TIT FOR TAT is extraordinarily robust. Perhaps it is not too much to hope that people can use the surrogate experience of the Computer Tournament to learn the value of reciprocity for their own Prisoner's Dilemma interactions.

Once the word gets out that reciprocity works, it becomes the thing to do. If you expect others to reciprocate your defections as well as your cooperations, you will be wise to avoid starting any trouble. Moreover, you will be wise to defect after someone else defects, showing that you will not be exploited. Thus you too will be wise to use a strategy based upon reciprocity. So will everyone else. In this manner the appreciation of the value of reciprocity becomes self-reinforcing. Once it gets going, it gets stronger and stronger.

This is the essence of the ratchet effect which was established in chapter 3: once cooperation based upon reciprocity gets established in a population, it cannot be overcome even by a cluster of individuals who try to exploit the others. The establishment of stable cooperation can take a long time if it is based upon blind forces of evolution, or it can happen rather quickly if its operation can be appreciated by intelligent players. The empirical and theoretical results of this book might help people see more clearly the opportunities for reciprocity latent in their world. Knowing the concepts that accounted for the results of the two rounds of the Computer Prisoner's Dilemma Tournament, and knowing the reasons and conditions for the success of reciprocity, might provide some additional foresight.

We might come to see more clearly that there is a lesson in the fact that TIT FOR TAT succeeds without doing better than anyone with whom it interacts. It succeeds by eliciting cooperation from others, not by defeating them. We are used to thinking about competitions in which there is only one winner, competitions such as football or chess. But the world is rarely like that. In a vast range of situations mutual cooperation can be better for *both* sides than mutual defection. The key to doing well lies not in overcoming others, but in eliciting their cooperation.

Today, the most important problems facing humanity are in the arena of international relations, where independent, egoistic nations face each other in a state of near anarchy. Many of these problems take the form of an iterated Prisoner's Dilemma. Examples can include arms races, nuclear proliferation, crisis bargaining, and military escalation. Of course, a realistic understanding of these problems would have to take into account many factors not incorporated into the simple Prisoner's Dilemma formulation, such as ideology, bureaucratic politics, commitments, coalitions, mediation, and leadership. Nevertheless, we can use all the insights we can get.

Robert Gilpin (1981, p. 205) points out that from the ancient Greeks to contemporary scholarship all political theory addresses one fundamental question: "How can the human race, whether for selfish or more cosmopolitan ends, understand and control the seemingly blind forces of history?" In the contemporary world this question has become especially acute because of the development of nuclear weapons.

The advice in chapter 6 to players of the Prisoner's Dilemma might serve as good advice to national leaders as well: don't be envious, don't be the first to defect, reciprocate both cooperation and defection, and don't be too clever. Likewise, the techniques discussed in chapter 7 for promoting cooperation in the Prisoner's Dilemma might also be useful in promoting cooperation in international politics.

The core of the problem of how to achieve rewards from cooperation is that trial and error in learning is slow and painful. The conditions may all be favorable for long-run developments, but we may not have the time to wait for blind processes to move us slowly toward mutually rewarding strategies based upon reciprocity. Perhaps if we understand the process better, we can use our foresight to speed up the evolution of cooperation.

APPENDIX A

Tournament Results

This appendix supplements chapter 2 by providing additional information about the two rounds of the Computer Prisoner's Dilemma Tournament. It provides information about the people who entered the tournament, about the entries themselves, and about how well each entry performed when matched up with each of the others. It also examines what would have happened under six major variations in the tournament, providing additional evidence for the robustness of TIT FOR TAT's success.

The first round of the tournament included fourteen entries plus RANDOM. The names of contestants and the scores of their decision rules are given in table 2. Each pair of rules was matched in five games of two hundred moves each. The tournament scores of each rule with each other rule is given in table 3. A description of each strategy is given in Axelrod (1980*a*) which is also the report made available to the entrants to the second round of the tournament.

TABLE 2
The Contestants: Round One

Rank	*Name*	*Discipline (if faculty)*	*Length of Program*	*Score*
1	Anatol Rapoport	Psychology	4	504.5
2	Nicholas Tideman & Paula Chieruzzi	Economics	41	500.4
3	Rudy Nydegger	Psychology	23	485.5
4	Bernard Grofman	Political Sci.	8	481.9
5	Martin Shubik	Economics	16	480.7
6	William Stein & Amnon Rapoport	Mathematics Psychology	50	477.8
7	James W. Friedman	Economics	13	473.4
8	Morton Davis	Mathematics	6	471.8
9	James Graaskamp		63	400.7
10	Leslie Downing	Psychology	33	390.6
11	Scott Feld	Sociology	6	327.6
12	Johann Joss	Mathematics	5	304.4
13	Gordon Tullock	Economics	18	300.5
14	Name withheld		77	282.2
15	RANDOM		5	276.3

The contestants in the second round are listed in table 4 along with some information about their programs. Each pair of rules was matched in five games of varying lengths, averaging 151 moves each. There were sixty-two entries plus RANDOM, so the tournament score matrix for the second round is a huge 63 by 63 matrix. It is so big that table 5 has to give it in compressed form (see table 5). The average score of each rule with each other rule is shown as a single digit according to the following code:

1: less than 100 points
2: 100-199.9 points (151 points is total mutual defection)
3: 200-299.9 points
4: 300-399.9 points
5: 400-452.9 points
6: exactly 453 points (total mutual cooperation)

TABLE 3

Tournament Scores: Round One

Player		Other Players															
		TIT FOR TAT	TIDE AND CHIER	NYDEG-GER	GROF-MAN	SHUBIK	STEIN AND RAP	FRIED-MAN	DAVIS	GRAAS-KAMP	DOWN-ING	FELD	JOSS	TUL-LOCK	(Name With-held)	RAN-DOM	Average Score
1.	TIT FOR TAT (Anatol Rapoport)	600	595	600	600	600	595	600	600	597	597	280	225	279	359	441	504
2.	TIDEMAN & CHIERUZZI	600	596	600	601	600	596	600	600	310	601	271	213	291	455	573	500
3.	NYDEGGER	600	595	600	600	600	595	600	600	433	158	354	374	347	368	464	486
4.	GROFMAN	600	595	600	600	600	594	600	600	376	309	280	236	305	426	507	482
5.	SHUBIK	600	595	600	600	600	595	600	600	348	271	274	272	265	448	543	481
6.	STEIN & RAPOPORT	600	596	600	602	600	596	600	600	319	200	252	249	280	480	592	478
7.	FRIEDMAN	600	595	600	600	600	595	600	600	307	207	235	213	263	489	598	473
8.	DAVIS	600	595	600	600	600	595	600	600	307	194	238	247	253	450	598	472
9.	GRAASKAMP	597	305	462	375	348	314	302	302	588	625	268	238	274	466	548	401
10.	DOWNING	597	591	398	289	261	215	202	239	555	202	436	540	243	487	604	391
11.	FELD	285	272	426	286	297	255	235	239	274	704	246	236	272	420	467	328
12.	JOSS	230	214	409	237	286	254	213	252	244	634	236	224	273	390	469	304
13.	TULLOCK	284	287	415	293	318	271	243	229	278	193	271	260	273	416	478	301
14.	(Name Withheld)	362	231	397	273	230	149	133	173	187	133	317	366	345	413	526	282
15.	RANDOM	442	142	407	313	219	141	108	137	189	102	360	416	419	300	450	276

TABLE 4
The Contestants: Round Two

Rank	*Name*	*Country (if not U.S.)*	*Discipline (if faculty)*	*Language (FORTRAN or BASIC)*	*Length of Program*[a]
1	Anatol Rapoport	Canada	Psychology	F	5
2	Danny C. Champion			F	16
3	Otto Bọrufsen	Norway		F	77
4	Rob Cave			F	20
5	William Adams			B	22
6	Jim Graaskamp & Ken Katzen			F	23
7	Herb Weiner			F	31
8	Paul D. Harrington			F	112
9	T. Nicolaus Tideman & P. Chieruzzi		Economics	F	38
10	Charles Kluepfel			B	59
11	Abraham Getzler			F	9
12	Francois Leyvraz	Switzerland		B	29
13	Edward White, Jr.			F	16
14	Graham Eatherley	Canada		F	12
15	Paul E. Black			F	22
16	Richard Hufford			F	45
17	Brian Yamauchi			B	32
18	John W. Colbert			F	63
19	Fred Mauk			F	63
20	Ray Mikkelson		Physics	B	27
21	Glenn Rowsam			F	36
22	Scott Appold			F	41
23	Gail Grisell			B	10
24	J. Maynard Smith	United Kingdom	Biology	F	9
25	Tom Almy			F	142
26	D. Ambuelh & K. Kickey			F	23
27	Craig Feathers			B	48
28	Bernard Grofman		Political Sci.	F	27
29	Johann Joss	Switzerland	Mathematics	B	74
30	Jonathan Pinkley			F	64
31	Rudy Nydegger		Psychology	F	23
32	Robert Pebley			B	13
33	Roger Falk & James Langsted			B	117
34	Nelson Weiderman		Computer Sci.	F	18
35	Robert Adams			B	43

TABLE 4 *(continued)*
The Contestants: Round Two

Rank	*Name*	*Country (if not U.S.)*	*Discipline (if faculty)*	*Language (FORTRAN or BASIC)*	*Length of Program*[a]
36	Robyn M. Dawes & Mark Batell		Psychology	F	29
37	George Lefevre			B	10
38	Stanley F. Quayle			F	44
39	R. D. Anderson			F	44
40	Leslie Downing		Psychology	F	33
41	George Zimmerman			F	36
42	Steve Newman			F	51
43	Martyn Jones	New Zealand		B	152
44	E.E.H. Shurmann			B	32
45	Henry Nussbacher			B	52
46	David Gladstein			F	28
47	Mark F. Batell			F	30
48	David A. Smith			B	23
49	Robert Leyland	New Zealand		B	52
50	Michael F. McGurrin			F	78
51	Howard R. Hollander			F	16
52	James W. Friedman		Economics	F	9
53	George Hufford			F	41
54	Rik Smoody			F	6
55	Scott Feld		Sociology	F	50
56	Gene Snodgrass			F	90
57	George Duisman			B	6
58	W. H. Robertson			F	54
59	Harold Rabbie			F	52
60	James E. Hall			F	31
61	Edward Friedland			F	84
62	RANDOM			F	(4)
63	Roger Hotz			B	14

[a] Length is given in terms of the number of internal statements in the FORTRAN version of the program. A conditional instruction is counted as two internal statements here, although it was counted as only one instruction in the report of the first round.

7: 453.1-499.9 points
8: 500-599.9 points
9: 600 or more points

Tournament Results

While table 5 can give some idea of why a given rule scored as it did, the amount of detail is overwhelming. Therefore, a more parsimonious method is needed to make sense of the results. Fortunately, stepwise regression provides such a method. It turns out that just five of the rules can be used to account very well for how well a given rule did with the entire set of 63. These five rules can thus be thought of as *representatives* of the full set in the sense that the scores a given rule gets with them can be used to predict the average score the rule gets over the full set.

TABLE 5

Tournament Scores: Round Two

	Other Players												
Player	1		11		21		31		41		51		61
1	66666	66566	66666	56556	66665	65656	66666	66656	66666	56555	56554	44452	442
2	66666	66566	66666	56556	66665	65656	66666	66656	66666	56555	56554	44552	442
3	66666	66566	66666	56556	66665	65656	66666	66656	66666	56555	56554	44443	452
4	66666	66566	66666	56556	66665	65656	66666	66656	66666	56555	56553	45542	352
5	66666	66566	66666	56546	66665	65656	66666	66656	66666	56545	36494	44542	442
6	66666	66566	66666	56556	66665	65656	66666	66656	66666	56555	46583	35232	353
7	66666	66566	66666	56546	66665	65656	66666	66656	66666	56555	56553	35272	253
8	55577	55555	55777	58558	75887	85455	45485	54888	58443	53758	53574	44543	452
9	66666	66566	66666	56556	66665	65656	66666	66656	66666	56455	56554	45232	272
10	66666	66566	66666	56546	66665	65656	66666	66656	66666	56554	56554	45342	352
11	66666	66566	66666	56536	66665	65656	66666	66656	66666	46534	56553	44552	342
12	66666	66566	66666	36546	66665	65646	66666	66656	66666	56555	56554	44553	242
13	66666	66466	66666	46556	66664	64656	66666	66646	66666	56544	56354	44552	442
14	66666	66566	66666	56556	66664	64656	66666	66646	66666	56535	56453	43533	432
15	66666	66566	66666	46556	66664	64656	66666	66646	66666	56544	56354	43532	232
16	55575	55555	54777	57557	75775	77757	43375	77777	47443	54757	42484	44222	452
17	66666	66466	66666	46556	66664	64656	66666	66646	66666	56534	56253	45533	253
18	57557	55555	55777	57547	75777	77557	35577	77777	77743	57555	51572	44553	142
19	55674	54564	35777	57557	75777	77757	43473	77777	47443	55757	53573	44572	453
20	66666	66466	66666	46556	66664	64656	66666	66646	66666	56534	56253	35532	252

TABLE 5 *(continued)*

Tournament Scores: Round Two

Player	*Other Players* 1		11		21		31		41		51		61
21	66666	66566	66666	46556	66664	64646	66666	66646	66666	56534	56353	35332	252
22	66666	66566	66666	56556	66664	65646	66666	66656	66666	36535	56353	44422	242
23	66666	66466	66666	46556	66663	64656	66666	66646	66666	56535	56252	33533	443
24	66666	66466	66666	46556	66663	64656	66666	66646	66666	36554	56454	43433	332
25	55575	55555	55878	58558	75885	85255	55384	38848	28433	52745	52583	45243	242
26	66666	66466	66666	46556	66664	64656	66666	66646	66666	56535	56353	34252	243
27	55575	55555	55777	57558	75874	75557	54373	75878	58434	53737	52353	44442	342
28	66666	66366	66666	46556	66663	65656	66666	66646	66666	56534	56353	35232	252
29	55575	55555	54777	57557	54775	75757	43473	77777	37343	53757	52474	55532	252
30	66666	66566	66666	46556	66664	64656	66666	66646	66666	46524	56352	34233	242
31	66666	66466	66666	36546	66667	67626	66666	66676	66666	46534	56773	44242	242
32	66666	66566	66666	36536	66665	64636	66666	66646	66666	26433	26573	45242	252
33	66666	66566	66666	36536	66663	63646	66666	66636	66666	36453	46394	35222	252
34	66666	66466	66666	46556	66663	65656	66666	66646	66666	36534	56253	33233	253
35	66666	66566	66666	56536	66664	63636	66666	66656	66666	26432	26493	45252	252
36	66666	66466	66666	46556	66663	64656	66666	66646	66666	26534	56253	35232	253
37	66666	66466	66666	46556	66664	64656	66666	66646	66666	26532	56353	35222	273
38	66666	66466	66666	46556	66663	64656	66666	66646	66666	36524	56272	33233	273
39	55555	55455	55778	48558	55883	84855	54384	44858	38335	52738	72373	35232	252
40	66666	66466	66666	46556	66663	64656	66666	66646	66666	36534	56272	33233	253
41	66666	66566	66666	46546	66663	65646	66666	66646	66666	26433	36373	45252	242
42	66666	66466	66666	46556	66663	64646	66666	66646	66666	36524	56252	33223	252
43	66666	66466	66666	46546	66664	64636	66666	66636	66666	26433	46383	44222	242
44	66666	66566	66666	36546	66663	63626	66666	66626	66666	46434	46393	45222	242
45	66666	66566	66666	36536	66664	65636	66666	66676	66666	26433	36373	35232	252
46	57557	55555	45757	56757	54597	55754	42392	22959	29233	52755	52574	44252	442
47	66666	66366	66666	46546	66663	63636	66666	66636	66666	26333	36393	35232	253
48	55575	55545	55777	57557	75774	74757	43373	77744	44433	53555	52454	43433	332
49	55554	55555	35785	54553	45575	34358	43533	43848	38343	53557	53593	45572	352
50	55575	55454	55757	47557	75575	54757	43372	72747	37343	54745	72354	43232	442
51	55573	45555	55777	47557	75775	75757	32372	77744	35433	54555	52454	43432	332
52	66666	66366	66666	36536	66663	63636	66666	66636	66666	26332	26392	35222	253
53	55564	55555	55375	33543	35385	34243	55324	32333	24332	52758	72593	44532	242
54	55552	35455	55777	37557	75774	75757	44171	77444	57314	52525	51152	23413	131
55	44434	33544	35575	43533	44454	33347	43433	33737	38343	43535	42494	45342	353

TABLE 5 *(continued)*
Tournament Scores: Round Two

Player	*Other Players* 1		11		21		31		41		51		61
56	55555	22544	45575	43542	34477	35348	44322	22424	45333	42725	52392	44233	442
57	44524	22712	44577	52442	24992	45114	42192	21929	39322	41829	81382	34923	442
58	45377	22433	55775	35647	35753	25247	42222	22222	23233	22542	52783	33533	253
59	55234	24532	55577	22552	24282	55224	43222	22222	33232	52838	82292	35853	252
60	22432	22742	27343	37522	33998	22235	42292	22828	28233	22722	32382	55982	542
61	44734	22734	34473	52742	23483	23222	42222	22222	23333	42724	72293	44223	253
62	44224	12212	45477	22422	24473	44212	32222	21123	32322	41724	51382	34223	142
63	33323	22533	34333	22522	23233	22233	42322	22222	23333	22333	32392	35232	252

Code:
1. less than 100 points
2. 100–199.9 points
3. 200–299.9 points
4. 300–399.9 points
5. 400–452.9 points
6. Exactly 453 points
7. 453.1–499.9 points
8. 500–599.9 points
9. 600 or more points

The formula for the predicted tournament score is:

$$T = 120.0 + (.202)S_6 + (.198)S_{30} + (.110)S_{35} + (.072)S_{46} + (.086)S_{27}$$

where T is the predicted tournament score of a rule, and S_j is the score which that rule gets with the jth rule.

This estimate of the tournament scores correlated with the actual tournament scores at $r = .979$, and $r^2 = .96$. This means that 96 percent of the variance in the tournament scores is explained by knowing a rule's performance with only the five representatives.

TIT FOR TAT's victory in the tournament can be explained by its good scores with all five of the representatives. Recall that 453 points is what is attained from unending mutual cooperation. TIT FOR TAT got the following scores with the five representatives: $S_6 = 453$; $S_{30} = 453$; $S_{35} = 453$; $S_{46} = 452$; and $S_{27} = 446$. Using these as the standard of comparison one can see how other rules did in the tournament by seeing how much worse (or better) they did with the five representatives compared to how TIT FOR TAT did with them. This display is provided in table 6 and will form the basis of the rest of the analysis of this round (see table).

TABLE 6
Performance of the Rules: Round Two

		Performance with Representatives (Points lost relative to TIT FOR TAT)					
Rank	*Tournament Score*	*Rule 6*	*Rev. State Transition (30)*	*Rule 35*	*Tester (46)*	*Tranquilizer (27)*	*Residual*
1	434.73	0	0	0	0	0	13.3
2	433.88	0	0	0	12.0	2.0	13.4
3	431.77	0	0	0	0	6.6	10.9
4	427.76	0	0	0	1.2	25.0	8.5
5	427.10	0	0	0	15.0	16.6	8.1
6	425.60	0	0	0	0	1.0	4.2
7	425.48	0	0	0	0	3.6	4.3
8	425.46	1.0	37.2	16.6	1.0	1.6	13.6
9	425.07	0	0	0	0	11.2	4.5
10	425.94	0	0	0	26.4	10.6	6.3
11	422.83	0	0	0	84.8	10.2	8.3
12	422.66	0	0	0	5.8	−1.2	1.5
13	419.67	0	0	0	27.0	61.4	5.4
14	418.77	0	0	0	0	50.4	1.6
15	414.11	0	0	0	9.4	52.0	−2.2
16	411.75	3.6	−26.8	41.2	3.4	−22.4	−11.5
17	411.59	0	0	0	4.0	61.4	−4.3
18	411.08	1.0	−2.0	−.8	7.0	−7.8	−10.9
19	410.45	3.0	−19.6	171.8	3.0	−14.2	3.5
20	410.31	0	0	0	18.0	68.0	−4.0
21	410.28	0	0	0	20.0	57.2	−4.9
22	408.55	0	0	0	154.6	31.8	.9
23	408.11	0	0	0	0	67.4	−7.6
24	407.79	0	0	0	224.6	56.0	7.2
25	407.01	1.0	2.2	113.4	15.0	33.6	2.5
26	406.95	0	0	0	0	59.6	−9.4
27	405.90	8.0	−18.6	227.8	5.6	14.0	8.9
28	403.97	0	0	0	3.0	1.4	−17.2
29	403.13	4.0	−24.8	245.0	4.0	−3.0	4.4

TABLE 6 *(continued)*
Performance of the Rules: Round Two

		Performance with Representatives (Points lost relative to TIT FOR TAT)					
Rank	*Tournament Score*	*Rule 6*	*Rev. State Transition (30)*	*Rule 35*	*Tester (46)*	*Tran-quilizer (27)*	*Residual*
30	402.90	0	0	0	74.0	54.4	−8.6
31	402.16	0	0	0	147.4	−10.0	−9.6
32	400.75	0	0	0	264.2	52.4	2.7
33	400.52	0	0	0	183.6	157.4	5.7
34	399.98	0	0	0	224.6	41.6	−1.9
35	399.60	0	0	0	291.0	204.8	16.5
36	399.31	0	0	0	288.0	61.4	3.7
37	398.13	0	0	0	294.0	58.4	2.7
38	397.70	0	0	0	224.6	84.8	−.4
39	397.66	1.0	2.6	54.4	2.0	46.6	−13.0
40	397.13	0	0	0	224.6	72.8	−2.0
41	395.33	0	0	0	289.0	−5.6	−6.0
42	394.02	0	0	0	224.6	74.0	−5.0
43	393.01	0	0	0	282.0	55.8	−3.5
44	392.54	0	0	0	151.4	159.2	−4.4
45	392.41	0	0	0	252.6	44.6	−7.2
46	390.89	1.0	73.0	292.0	1.0	−.4	16.1
47	389.44	0	0	0	291.0	156.8	2.2
48	388.92	7.8	−15.6	216.0	29.8	55.2	−3.5
49	385.00	2.0	−90.0	189.0	2.8	101.0	−24.3
50	383.17	1.0	−38.4	278.0	1.0	61.8	−9.9
51	380.95	135.6	−22.0	265.4	26.8	29.8	16.1
52	380.49	0	0	0	294.0	205.2	−2.3
53	344.17	1.0	199.4	117.2	3.0	88.4	−17.0
54	342.89	167.6	−30.8	385.0	42.4	29.4	−3.1
55	327.64	241.0	−32.6	230.2	102.2	181.6	−3.4
56	326.94	305.0	−74.4	285.2	73.4	42.0	−7.5
57	309.03	334.8	74.0	270.2	73.0	42.2	8.4
58	304.62	274.0	−6.4	290.4	294.0	6.0	−9.3
59	303.52	302.0	142.2	271.4	13.0	−1.0	1.8
60	296.89	293.0	34.2	292.2	291.0	286.0	18.8
61	277.70	277.0	262.4	293.0	76.0	178.8	17.0
62	237.22	359.2	261.8	286.0	114.4	90.2	−12.6
63	220.50	311.6	249.0	293.6	259.0	254.0	−16.2

Also provided in table 6 are actual tournament scores for each rule and the residual that is the difference between the actual tournament score and the predicted tournament score. Notice that while the tournament scores cover a range of several hundred points, the residuals are usually smaller than 10 points, indicating again how well the five representatives account for the overall performance of the rules. Another interesting feature of the residuals is that the top-ranking rules tend to have the largest positive residuals indicating that they do better than most of the rules on the limited aspects of the tournament which are not accounted for by the five representatives.

The representatives can now be used to help answer the central questions of what worked and why.

Table 6 shows the pattern very clearly in the scores with the five representatives. The first three representatives are themselves nice. All of the nice rules got 453 points with each of these three, so the nice rules lost no points compared to how first-placed TIT FOR TAT did with them. The rules which were not nice generally did not do as well as TIT FOR TAT did with these first three representatives, as shown by the predominance of positive over negative numbers in these three columns of table 6.

To give an example, the best of the rules which was not nice was submitted by Paul Harrington and ranked eighth. This rule is a variant of TIT FOR TAT which has a check for RANDOM, and a way of getting out of alternating defections (echo effects), and also a method of seeing what it can get away with. It always defects on move 37 and with increasing probability after that unless the other player defects immediately after one of these defections, in which case it no longer defects randomly. It did not do as well as TIT FOR TAT with any of the five representatives, but it suffered most from the second representative. With that entry it got 37.2 points less than TIT FOR TAT did. This second representative is REVISED STATE TRANSITION, modified from the supplementary rule of round one and submitted in round two by Jonathan Pinkley. REVISED STATE TRANSITION models the other player as a one-step Markov process. It makes its own choice so as to maximize its own long-term payoff on the assumption that this model is correct. As Harrington's rule defected more and more, the REVISED STATE TRANSITION rule kept a running estimate of the probability that the other would cooperate after each of the four possible outcomes. Eventually REVISED STATE TRANSITION determined that it did not pay to cooperate after the other exploited it, and soon thereafter it also determined that it did not even pay to cooperate after a mutual cooperation.[1]

So even if the other rule is willing to accept some defections, once the limit of its tolerance is reached it is hard to convince it that one's ways have been mended. While some of the other rules that were not nice did in fact manage to do better than TIT FOR TAT with REVISED STATE TRANSITION, these rules tended to do much worse with some of the other representatives.

The five representatives can be used not only to analyze the results of the second round of the tournament, but also to construct hypothetical variants of the tournament. This is done by assigning a different relative weight to each of the types of rules participating. The five representatives can each be thought of as having a large constituency. Together with the unrepresented constituency of the residuals, these five constituencies fully account for the performance of each rule in the tournament. The use of representatives allows an investigation to see what would have happened if one of the constituencies had been much larger than it actually was. To be specific, the hypothetical tournaments are those which would have resulted if a given constituency had been five times as large as it actually was. Since there are six constituencies, this provides six hypothetical tournaments. Each of these hypothetical tournaments represents a substantial variation on the original tournament because it quintuples the size of one or another of the six constituencies. And each represents a different kind of variation since each is based on magnifying the effect of a different aspect of a rule's environment.[2]

In fact, the scores in these hypothetical tournaments correlate fairly well with the scores in the original tournament. If the residuals were five times as large as they actually were, the tournament scores would still have a correlation of .82 with the scores in the actual tournament. And if the constituency of any of the five representatives were made five times as large as it actually was, the tournament scores would still be correlated from .90 to .96 with the tournament scores of the actual second round. This means that the overall results would have been fairly stable even if the distribution of entries by types of program had been quite different from what it actually was. Thus the overall results of the second round are quite robust.

But moving from the tournament as a whole to the identity of the winner, one can also ask how TIT FOR TAT would have done in these six hypothetical tournaments. The answer is that it would still have come in first place in five of the six hypothetical tournaments. This is a very strong result since it shows that TIT FOR TAT would still have been the best rule of those submitted under very wide variations in the environment it had to face.

The one exception to TIT FOR TAT's success in the

hypothetical tournaments is a very interesting one. Had the constituency of the REVISED STATE TRANSITION rule been five times as large as it actually was, TIT FOR TAT would have come in second. First place would have been won by a rule which ranked only forty-ninth in the actual tournament. This rule was submitted by Robert Leyland of Auckland, New Zealand. Its motivation is similar to TRANQUILIZER's in that it starts off cooperatively but then sees how much it can get away with. As can be seen from table 6, Leyland's rule came in fourty-ninth largely because it did so poorly with the third representative and with TRANQUILIZER. But it did do 90 points better than TIT FOR TAT with REVISED STATE TRANSITION, since that rule was quite well taken in by the early cooperations. If the constituency of the REVISED STATE TRANSITION representative had been five times as large as it actually was, Leyland's rule would actually have done better than TIT FOR TAT or any other submitted rule in the tournament as a whole.

The fact that TIT FOR TAT won five of the six major variants of the tournament and came in second in the sixth shows that TIT FOR TAT's victory was very robust indeed.

A P P E N D I X B

Proofs of the Theoretical Propositions

THIS APPENDIX reviews the theoretical propositions and provides the proofs of those not already given in the text. It also provides the theoretical result that characterizes all collectively stable strategies.

The Prisoner's Dilemma game is defined as a two-player game in which each player can either cooperate (*C*) or defect (*D*). If both cooperate, both get the reward *R*. If both defect, both get the punishment *P*. If one cooperates and the other defects, the first gets the sucker's payoff, *S*, and the other gets the temptation, *T*. The payoffs are ordered $T > R > P > S$, and satisfy $R > (T+S)/2$. The game matrix is shown with representative values in figure 1 of chapter 1 (see page 8). In the iterated Prisoner's Dilemma, each move is worth less than the move before, by a factor of w, where $0 < w < 1$. Therefore in the iterated game, the cumulative payoff to either of two players who always cooperate with each other is $R + wR + w^2R \ldots = R/(1-w)$.

A strategy is a function from the entire history of the game so far to a probability of cooperating on the current move. A typical strategy is TIT FOR TAT, which is certain to cooperate on the first move, and then always does what the other player did on the previous move. In general, the value (or score) of strategy A when interacting with strategy B is represented by $V(A|B)$. Strategy A is said to invade a population consisting of players using strategy B if $V(A|B) > V(B|B)$. If no strategy exists which can invade B, then B is said to be collectively stable.

The first proposition gives the sad news that if the future is important enough, there is no one best strategy in the iterated Prisoner's Dilemma.

Proposition 1. If the discount parameter, w, is sufficiently high, there is no best strategy independent of the strategy used by the other player.

The proof is given in chapter 1.

The second proposition says that if everyone is using TIT FOR TAT, and the future is important enough, then no one can do any better by switching to another strategy.

Proposition 2. TIT FOR TAT is collectively stable if and only if w is at least as great as the larger of $(T-R)/(T-P)$ and $(T-R)/(R-S)$.

Proof. First this proposition is shown to be equivalent to saying that TIT FOR TAT is collectively stable if and only if it is invadable neither by ALL D nor the strategy which alternates defection and cooperation. After proving the two formulations are equivalent, both implications of the second formulation are proved.

To say that ALL D cannot invade TIT FOR TAT means that $V(\text{ALL D}|\text{TFT}) \leq V(\text{TFT}|\text{TFT})$. When ALL D meets TIT FOR TAT, it gets T on the first move and P thereafter, making $V(\text{ALL D}|\text{TFT}) = T + wP/(1-w)$. Since TIT FOR TAT always cooperates with its twin, $V(\text{TFT}|\text{TFT}) = R + wR + w^2R \ldots = R/(1-w)$. Thus ALL D cannot invade TIT FOR TAT when $T + wP/(1-w) \leq R/(1-w)$, or $T(1-w) + wP \leq R$, or $T - R \leq w(T - P)$ or $w \geq (T-R)/(T-P)$. Similarly, to say that alternation of D and C cannot invade TIT FOR TAT means that $(T+wS)/(1-w^2) \leq R/(1-w)$, or $(T-R)/(R-S) \leq w$. Thus $w \geq (T-R)/(T-P)$ and $w \geq (T-R)/(R-S)$ is equivalent to saying that TIT FOR TAT is invadable by neither ALL D nor the strategy which alternates defection and cooperation. This shows that the two formulations are equivalent.

Now both of the implications of the second formulation will be proved. One implication is established by the simple observation that if TIT FOR TAT is a collectively stable strategy, then no rule can invade, and hence neither can the two specified rules. The other implication to be proved is that if neither ALL D nor Alternation of D and C can invade TIT FOR TAT, then no strategy can. TIT FOR TAT has only two states, depending on what the other player did the previous move (on the first move it assumes, in effect, that the other player has just cooperated). Thus if A is interacting with TIT FOR TAT, the best which any strategy, A, can do after choosing D is to choose C or D. Similarly, the best A can do after choosing D is to choose C or D. This leaves four possibilities for the best A can do with TIT FOR TAT: repeated sequences of *CC, CD, DC,* or *DD.* The first does the same as TIT FOR TAT does with another TIT FOR TAT. The second cannot do better than both the first and the third. This implies if the third and fourth possibilities cannot invade TIT FOR TAT, then no strategy can. These two are equivalent, respectively, to Alternation of *D* and *C,* and ALL D. Thus if neither of these two can invade TIT FOR TAT, no rule can, and TIT FOR TAT is a collectively stable strategy. This completes the proof.

Having proven when TIT FOR TAT is collectively stable, the next big step is to characterize all collectively stable strategies. The characterization of all collectively stable strategies is based on the idea that invasion can be prevented if the common rule makes the potential invader worse off than if it had just followed the common strategy. Rule *B* can prevent invasion by rule *A* if *B* can be sure that no matter what *A* does later, *B* will hold *A*'s total score low enough. This leads to the following useful definition: *B* has a *secure position* over *A* on move n if no matter what *A* does from move n onward, $V(A|B) \leq V(B|B)$, assuming that *B* defects from move n onward. Let $V_n(A|B)$ represent *A*'s discounted cumulative score in the moves before move n. Then another way of saying that *B* has a secure position over *A* on move n is that

$$V_n(A|B) + w^{n-1}P/(1-w) \leq V(B|B),$$

since the best *A* can do from move n onward if *B* defects is get *P* each time.

The theorem which follows embodies the advice that if you want to employ a collectively stable strategy, you should only cooperate when you can afford an exploitation by the other side and still retain your secure position.

The Characterization Theorem. *B* is a collectively stable strategy if and only if B defects on move n whenever the other player's cumulative score so far is too great, specifically when $V_n(A|B) > V(B|B) - w^{n-1}[T + wP/(1-w)]$.

The proof is given in Axelrod (1981).

The Characterization Theorem is "policy relevant" in the abstract sense that it specifies what a strategy, B, has to do at any point in time as a function of the previous history of the interaction in order for B to be a collectively stable strategy.[1] It is a complete characterization because this requirement is both a necessary and a sufficient condition for strategy B to be collectively stable.

Two additional consequences about collectively stable strategies can be seen from the theorem. First, as long as the other player has not accumulated too great a score, a strategy has the flexibility to either cooperate or defect and still be collectively stable. This flexibility explains why there are typically many strategies which are collectively stable. The second consequence is that a nice rule (one which will never defect first) has the most flexibility since it has the highest possible score when playing an identical rule. Put another way, nice rules can afford to be more generous than other rules with potential invaders because nice rules do so well with each other.

Proposition 2 demonstrated that TIT FOR TAT was collectively stable only when the future was important enough. The next proposition uses the Characterization Theorem to show that this conclusion is actually quite general. In fact it holds true for any strategy which may be the first to cooperate.

Proposition 3. Any strategy, B, which may be the first to cooperate can be collectively stable only when w is sufficiently large.

Proof. If B cooperates on the first move, $V(\text{ALL D}|B)$

$\geq T + wP/(1-w)$. But for any B, $R/(1-w) \geq V(B|B)$ since R is the best B can do with another B by the assumptions of the Prisoner's Dilemma that $R>P$ and $R>(S+T)/2$. Therefore $V(\text{ALL D}|B) > V(B|B)$ is so whenever $T + wP/(1-w) > R/(1-w)$. This implies that ALL D invades a B which cooperates on the first move whenever $w<(T-R)/(T-P)$. If B has a positive chance of cooperating on the first move, then the gain of $V(\text{ALL D}|B)$ over $V_1(B|B)$ can only be nullified if w is sufficiently large. Likewise, if B will not be the first to cooperate until move n, $V_n(\text{ALL D}|B) = V_n(B|B)$ and the gain of V_{n+1} (ALL D$|B$) over $V_{n+1}(B|B)$ can only be nullified if w is sufficiently large.

As noted earlier, a consequence of Characterization Theorem is that a nice rule has the most flexibility.

The flexibility of a nice rule is not unlimited, however, as shown by the following theorem. In fact, a nice rule must be *provoked* by the very first defection of the other player, i.e., on some later move the rule must have a finite chance of retaliating with a defection of its own.

Proposition 4. For a nice strategy to be collectively stable, it must be provoked by the very first defection of the other player.

Proof. If a nice strategy were not provoked by a defection on move n, then it would not be collectively stable because it could be invaded by a rule which defected only on move n.

There is one strategy which is *always* collectively stable, that is regardless of the value of w or the payoff parameters T, R, P, and S. This is ALL D, the rule which defects no matter what.

Proposition 5. ALL D is always collectively stable.

Proof. ALL D is always collectively stable because it

always defects and hence it defects whenever required by the condition of the Characterization Theorem.

This says that a world of "meanies" can resist invasion by anyone using any other strategy—provided that the newcomers arrive one at a time. So in order for the evolution of cooperation to get going, the newcomers must arrive in clusters. Assuming the new A's are rare relative to the established B's, by clustering together the A's can provide a significant part of each other's environment, but a negligible part of the B's environment. Thus, one says that a *p-cluster of A invades B* if $pV(A|A) + (1\text{-}p)V(B|B) > V(B|B)$, where p is the proportion of interactions by a player using strategy A with another such player. Solving for p, this means that invasion is possible if the newcomers interact enough with each other.

Notice that this assumes pairing in the interactions is not random. With random pairing, an A would rarely meet another A. Instead, the clustering concept treats the case in which the A's are a trivial part of the environment of the B's, but a nontrivial part of the environment of the other A's.

Chapter 3 gave numerical examples to show that invasion by clusters can, in fact, be surprisingly easy. For example, with the standard parametric values of $T=5$, $R=3$, $P=1$, and $S=0$, and with $w=.9$, a cluster of TIT FOR TATs can invade a population of meanies even if only 5 percent of their interactions are with other members of the cluster.

One might also ask what happens when the newcomers grow in numbers so that they are no longer a negligible part of the environment of the natives. As the proportion of newcomers grows, their need to avoid random mixing declines. Assuming completely random mixing with q percent newcomers, the newcomers will do better than the natives when $qV(A|A) + (1-q)V(A|B) > qV(B|A) + (1-q)V(B|B)$. Using the case of TIT FOR TAT invading ALL D, and using the standard payoff values, gives the modest requirement that $q > 1/17$. The newcomers can therefore thrive in a random mix as soon as they become a few percent of the entire population.

The chronological story begins with a cluster that is a negligible proportion of the whole population. It can get established provided its members have even a small chance, p, of meeting with each other. Then, once the new strategy thrives, it becomes less dependent on nonrandom mixing. Finally, when its numbers become even a few percent of the entire population, q, it can continue to thrive even with completely random mixing.

The next result shows which strategies are the most efficient at invading ALL D with the least amount of clustering. These are the strategies which are best able to discriminate between themselves and ALL D. A strategy is *maximally discriminating* if it will eventually cooperate even if the other has never cooperated yet, and once it cooperates it will never cooperate again with ALL D but will always cooperate with another player using the same strategy.

Proposition 6. The strategies which can invade ALL D in a cluster with the smallest value of p are those which are maximally discriminating, such as TIT FOR TAT.

Proof. To be able to invade ALL D, a rule must have a positive chance of cooperating first. Stochastic cooperation is not as good as deterministic cooperation with another player using the same rule since stochastic cooperation yields equal probability of S and T, and $(S+T)/2<R$ in the Prisoner's Dilemma. Therefore, a strategy which can

invade with the smallest p must cooperate first on some move, n, even if the other player has never cooperated yet. The definition of what it takes for a p-cluster of A to invade B implies that the rules which invade B=ALL D with the lowest value of p are those which have the lowest value of p^*, where $p^* = [V(B|B) - V(A|B)]/[V(A|A) - V(A|B)]$. The value of p^* is minimized when $V(A|A)$ and $V(A|B)$ are maximized (subject to the constraint that A cooperates for the first time on move n) since $V(A|A) > V(B|B) > V(A|B)$. $V(A|A)$ and $V(A|B)$ are maximized subject to this constraint if and only if A is a maximally discriminating rule. (Incidentally, it does not matter for the minimal value of p when A starts to cooperate.) TIT FOR TAT is such a strategy because it always cooperates for $n=1$, it cooperates only once with ALL D, and it always cooperates with another TIT FOR TAT.

The next proposition demonstrates that nice rules (those which never defect first) are actually better able than other rules to protect themselves from invasion by a cluster.

Proposition 7. If a nice strategy cannot be invaded by a single individual, it cannot be invaded by any cluster of individuals either.

Proof. For a cluster of rule A to invade a population of rule B, there must be a $p \leq 1$ such that $pV(A|A) + (1-p)V(A|B) > V(B|B)$. But if B is nice, then $V(A|A) \leq V(B|B)$. This is so because $V(B|B) = R/(1-w)$, which is the largest value attainable when the other player is using the same strategy. It is the largest value since $R > (S+T)/2$. Since $V(A|A) \leq V(B|B)$, A can invade as a cluster only if $V(A|B) > V(B|B)$. But that is equivalent to A invading as an individual.

The final result deals with a territorial system in which players interact only with their neighbors. In each generation, each player receives a success score which is the average of its performance with its neighbors. Then if a player has one or more neighbors who are more successful, the player converts to the strategy of the most successful of them (or picks randomly among the best in case of a tie among the most successful neighbors).

The concepts of invasion and stability are extended to territorial systems in the following manner. Suppose that a single individual using strategy *A* is introduced into one of the locations of a population where everyone else is using strategy *B*. One says that *A territorially invades B* if every location in the territory will eventually convert to strategy *A*. Then one can say that strategy *B* is *territorially stable* if no strategy can territorially invade it.

This leads to a strong result.

Proposition 8. If a rule is collectively stable, it is territorially stable.

The proof is given in chapter 8 for territorial systems based on a rectangular grid. The proof immediately generalizes to any territorial system which is not too highly interconnected. Specifically, it applies to any system which has the property that for every point, there exists a neighbor of a neighbor which is not a neighbor of the original point.

This demonstrates that protection from invasion is at least as easy in a territorial system as in a freely mixing system. An important implication is that mutual cooperation can be sustained in a (not too highly connected) territorial system at least as easily as it can be in a freely mixing system.

NOTES

Chapter 1. *The Problem of Cooperation*

1. For useful illustrations of these applications to international politics, see the following sources: the security dilemma (Jervis 1978), arms competition and disarmament (Rapoport 1960), alliance competition (Snyder 1971), tariff negotiations (Evans 1971), taxation of multinational firms (Laver 1977), and communal conflict in Cyprus (Lumsden 1973).

2. The Prisoner's Dilemma game was invented in about 1950 by Merrill Flood and Melvin Dresher, and formalized by A. W. Tucker shortly thereafter.

3. The situations that involve more than pairwise interaction can be modeled with the more complex *n*-person Prisoner's Dilemma (Olson 1965; G. Hardin 1968; Schelling 1973; Dawes 1980; R. Hardin 1982). The principal application is to the provision of collective goods. It is possible that the results from pairwise interactions will help suggest how to undertake a deeper analysis of the *n*-person case as well, but that must wait. For a parallel treatment of the two-person and *n*-person cases, see Taylor (1976, pp. 29–62).

4. The value received from always defecting when the other is playing TIT FOR TAT is:

$$\begin{aligned} V(\text{ALL D} \mid \text{TFT}) &= T + wP + w^2P + w^3P \ldots \\ &= T + wP(1+w+w^2 \ldots) \\ &= T + wP/(1-w). \end{aligned}$$

5. If the other player is using a strategy of permanent retaliation, you are better off always cooperating than ever defecting when $R/(1-w) > T + wP/(1-w)$ or $w > (T-R)/(T-P)$.

6. This means that the utilities need only be measured as an interval scale. Using an interval scale means that the representation of the payoffs may be altered with any positive linear transformation and still be the same, just as temperature is equivalent whether measured in Fahrenheit or Centigrade.

7. For the implications of not assuming deliberate choice in an evolutionary model of economic change, see Nelson and Winter (1982).

Chapter 2. *The Success of TIT FOR TAT in Computer Tournaments*

1. The second round of the tournament used a variable game length, as described in the text.
2. This is a broader definition of forgiveness than the one used by Rapoport and Chammah (1965, pp. 72–73), which is the probability of cooperation on the move after receiving the sucker's payoff, *S*.
3. In the five games between them, the average scores were 225 for TIT FOR TAT and 230 for JOSS.
4. In the environment of the 15 rules of the tournament, REVISED DOWNING averages 542 points. This compares to TIT FOR TAT, which won with 504 points. TIT FOR TWO TATS averages 532 in the same environment, and LOOK AHEAD averages 520 points.
5. This probability of ending the game at each move was chosen so that the expected median length of a game would be 200 moves. In practice, each pair of players was matched five times, and the lengths of these five games were determined once and for all by drawing a random sample. The resulting random sample from the implied distribution specified that the five games for each pair of players would be of lengths 63, 77, 151, 156, and 308 moves. Thus the average length of a game turned out to be somewhat shorter than expected at 151 moves.
6. This reproduction process creates a simulated second generation of the tournament in which the average score achieved by a rule is the *weighted* average of its score with each of the rules, where the weights are proportional to the success of the other rules in the initial generation.
7. This simulation of future rounds of the tournament is done by calculating the weighted average of the scores of a given rule with all other rules, where the weights are the numbers of the other rules which exist in the current generation. The numbers of a given rule in the next generation are then taken to be proportional to the product of its numbers in the current generation and its score in the current generation. This procedure assumes cardinal measurement of the payoff matrix. It is the only instance in this book where the payoff numbers are given a cardinal, rather than merely interval, interpretation.

Chapter 3. *The Chronology of Cooperation*

1. Those familiar with the concepts of game theory will recognize this definition of a collectively stable strategy as a strategy that is in Nash equilibrium with itself. My definitions of invasion and collective stability are slightly different from Maynard Smith's (1974) definitions of invasion and evolutionary stability. His definition of invasion allows a newcomer meeting a native to get exactly the *same* score as a native meeting a native, provided that a native meeting a newcomer does better than a newcomer meeting another newcomer. I have used the new definitions to simplify the proofs and to highlight the difference between the effect of a single mutant and the effect of a small number of mutants. Any rule which is evolutionarily stable is also collectively stable. For a nice rule (one which will never be the first to defect), the definitions are equivalent. All propositions in the text remain true if "evolutionary stability" is substituted for "collective stability" with the exception of the Characterization Theorem of Appendix B, where the characterization is necessary but no longer sufficient.

2. Collective stability can also be interpreted in terms of a commitment by one player, rather than the stability of a whole population. Suppose a player is committed to using a given strategy. Then another player can do no better than to use this same strategy if and only if the strategy is collectively stable.

3. The approach of limiting the situation was used in a variety of games by Hamilton (1967), and the approach of limiting the strategies was used by Maynard Smith and Price (1973), Maynard Smith (1978) and Taylor (1976). For related results on the potential stability of cooperative behavior see Luce and Raiffa (1957, p. 102), Kurz (1977), and Hirshleifer (1978).

4. In particular, the critical value of w to make TIT FOR TAT collectively stable is the larger of $(T-R)/(T-P)$ and $(T-R)/(R-S)$. As already seen in chapter 1, the score of ALL D when playing TIT FOR TAT is $T + wP + w^2P \ldots = T + wP/(1-w)$. This will be no better than the population average of $R/(1-w)$ when $w \geq (T-R)/(T-P)$. Similarly, the alternation of D and C when playing TIT FOR TAT will get $T + wS + w^2T + w^3S \ldots = (T + wS)(1 + w^2 + w^4 \ldots) = (T + wS)/(1-w^2)$. This will be no better than the population average of $R/(1-w)$ when $w \geq (T-R)/(R-S)$. For the full proof, see Appendix B.

5. A countervailing consideration is that a legislator in electoral trouble may receive help from friendly colleagues who wish to increase the chances of reelection of someone who has proven in the past to be cooperative, trustworthy, and effective.

6. In analyzing the tournament results, a concept related to provocability was found to be useful. This is a *retaliatory* rule, namely a rule which defects immediately after an "uncalled for" defection by the other player. The concept of provocability does not require certainty of a response, nor does it require an immediate response. The concept of a retaliatory rule requires both.

7. TIT FOR TAT playing with ALL D gets $S + wP + w^2P \ldots$ which is $S + wP/(1-w) = 0 + (.9 \times 1)/.1 = 9$ points.

8. The TIT FOR TAT players in a cluster will do better than the meanies if

$$30p + 9(1-p) > 10$$

or

$$21p + 9 > 10$$

or

$$21p > 1$$

or

$$p > 1/21.$$

This calculation ignores the negligible increase in the score of a typical native due to the presence of the tiny cluster of newcomers. For more details see Appendix A.

9. For details see Appendix B.

Chapter 4. *The Live-and-Let-Live System in Trench Warfare in World War I*

1. Ashworth (1980, pp. 171–75) estimates that the live-and-let-live system occurred in about one-third of all trench tours by British divisions.

Chapter 5. *The Evolution of Cooperation in Biological Systems (with William D. Hamilton)*

1. For more on the individualistic emphasis of Darwin's theory see Williams (1966) and Hamilton (1975). For the best recent case for effective selection at group levels and for altruism based on genetic correlation of unrelated players, see D. S. Wilson (1979).

2. On kinship theory, see Hamilton (1964). On reciprocity theory, see Trivers (1971), Chase (1980), Fagen (1980), and Boorman and Levitt (1980).

3. Caullery (1952) gives examples of antagonism in orchid-fungus and lichen symbioses. For the example of wasp-ant symbiosis, see Hamilton (1972).

4. There are many other patterns of interaction besides the Prisoner's Dilemma which allow gains from cooperation. See for example the model of combat between members of the same species in Maynard Smith and Price (1973).

5. For more on defection in evolution, see Hamilton (1971). Fagen (1980) shows some conditions for single encounters where defection is not the solution.

6. The parameter w can also take into account the discount rate between interactions, as explained in chapter 1.

7. This definition of an evolutionarily stable strategy (ESS) is due to Maynard Smith and Price (1973). For the closely related concept of collective stability see chapter 3, especially the first footnote.

8. Whether choice is simultaneous or sequential, cooperation on a tit-for-tat basis is evolutionarily stable if and only if w is sufficiently high. In the case of sequential moves, suppose there is a fixed chance, q, that a given player of the pair will be the next one to need help. The critical value of w can be shown to be the minimum of the two sides' value of $A/q(A+B)$ where A is the cost of giving assistance and B is the benefit of assistance when received. For examples of such assistance see Thompson (1980).

9. Yonge (1934) gives other examples of invertebrates with unicellular algae.

10. As specified in proposition 2 of chapter 3, the threshold for the stability of TIT FOR TAT is the maximum of $(T-R)/(T-P)$ and $(T-R)/(R-S)$.

11. See also Eshel (1977) for a related possible implication of multiclonal infection. For recent evidence on the ability of a virus to use a conditional strategy, see Ptashne, Johnson, and Pabo (1982).

Chapter 6. *How to Choose Effectively*

1. Behr (1981) uses this standard to recompute the scores of the first round of the Computer Prisoner's Dilemma Tournament. He points out that in some settings players seek to maximize their relative, rather than absolute, gain. Under this interpretation, however, the game is no longer a Prisoner's Dilemma, but is instead a zero-sum game with ALL D being the one and only dominant strategy for any value of w.

2. These two standards of player comparison can be expressed in formal terms, using the expression $V(A|B)$ to represent the expected value of strategy A when interacting with strategy B. The common mistake people make is to compare $V(A|B)$ to $V(B|A)$, and then to try to make sure they are doing better than the other player. The proper goal of the game, as reflected in the structure of the tournament, is to attain the highest possible score over all your different interac-

tions. This means maximizing the average of $V(A|B)$ over all the B's to be encountered. When meeting a player using a particular strategy, B, a good standard of comparison is whether you are doing as well as you could, given that the other player is using strategy B. What you should compare to the performance of your strategy, A, is the performance of another strategy, A′, when playing with this same B. This means comparing $V(A|B)$ with $V(A'|B)$. Overall, what you want is the strategy which does best on average with all the other B's you are going to encounter.

3. For more accounts of relationships between Gypsies and non-Gypsies, see also Kenrick and Puxon (1972), Quintana and Floyd (1972), Acton (1974), and Sway (1980).

4. This example of the effectiveness of clustering was based on $w=.9$, $T=5$, $R=3$, $P=1$, and $S=0$.

Chapter 7. *How to Promote Cooperation*

1. The score of ALL D when playing with TIT FOR TAT is $T + wP + w^2P \ldots$ which is $T + wP(1+w+ \ldots)$ which is $T + wP/(1-w)$. Numerically, this is $5 + .9 \times 1/.1 = 14$ points.

2. Alternating defection and cooperation when the other player is using TIT FOR TAT gives a score of $T + wS + w^2T + w^3S \ldots$ which can be simplified by grouping pairs of terms together and getting $(T + wS)(1 + w^2 + w^4 + w^6 \ldots)$. This is $(T + wS)/(1-w^2)$ or $(5 + .0)/(1 - .9 \times .9) = 26.3$.

3. Proposition 2 gives the relationship among the parameters which is needed for stability. A different approach would be to minimize the conflict of interest in the payoff matrix itself. To do this, the goal would be to decrease T and P, and to increase R and S (Rapoport and Chammah 1965, pp. 35–38; Axelrod 1970, pp. 65–70).

4. Altruism has generated an extensive literature in the social sciences. In public affairs people often act in socially responsible ways, for example by recycling used bottles (Tucker 1978) or donating blood (Titmuss 1971). In fact, altruism is so hard to explain in public affairs that a political scientist (Margolis 1982) has suggested that people might have one utility function for private affairs and another for public affairs. Among economists there is an interest in how to account for apparently altruistic deeds, and how to model the effects of altruism (e.g. Becker 1976; Kurz 1977; Hirshleifer 1977; and Wintrobe 1981). Among psychologists, the roots of altruism have been investigated experimentally. (For a review, see Schwartz 1977.) Game theorists have studied the theoretical implications of utility interaction (e.g. Valavanis 1958 and Fitzgerald 1975). Legal scholars have also investigated the conditions under which there is actually a legal obligation to rescue someone in trouble (Landes and Posner 1978*a* and 1978*b*).

5. Likewise, bacteria could not undertake complex feats of information processing about the history of the game so far, but they could presumably react to simple features of the past such as whether the environment has been more or less benign recently.

Chapter 8. *The Social Structure of Cooperation*

1. In the terminology of market signaling, this is called an index (Spence 1974).

2. Being meek gives $S + wR + w^2S + w^3R \ldots = (S+wR)/(1-w^2)$. If you revolt, you might as well defect all the time, which gives $P + wP + w^2P + w^3P \ldots = (P+wP)/(1-w^2)$. So there is no incentive to revolt whenever $(S+wR)/(1-w^2) > (P+wP)/(1-w^2)$. This is so when $S+wR > P+wP$, or $w > (P-S)/(R-P)$. So when w is large enough, there is no incentive to revolt. For the illustrative values of $S=0$, $P=1$, and $R=3$, it doesn't pay to revolt when w is greater than 1/2.

3. The concept of an evolutionarily stable strategy is similar to the concept of a collective stable strategy, and in the case of a nice rule, it is equivalent as explained in the first footnote of chapter 3.

4. With these values and $w = 1/3$, the territorial system gives $D_n > T_{n-1} > D_{n-1}$ except for $D_3 > T_4$. Here D_n is the score of an ALL D with n TIT FOR TAT neighbors, and T_n is the score of a TIT FOR TAT with n TIT FOR TAT neighbors. For example, $D_4 = V(\text{ALL D}|\text{TIT FOR TAT}) = T + wP/(1-w) = 56 + (1/3)(6)/(2/3) = 59$.

5. Some interesting possibilities which await examination are the following:

1. The ending of the interaction can depend on the history of the interaction. For example, it might depend on how well the players are doing. An unsuccessful player is more likely to die, go bankrupt, or seek another partner. An implication is that it might not pay to exploit a player who will not or cannot retaliate. The reason is that you shouldn't kill the goose that is laying golden eggs.

2. The game need not be an iterated Prisoner's Dilemma. For example, it could be an iterated game of chicken in which the worst outcome was mutual defection, as in crisis bargaining or labor strikes (Jervis 1978). For results on the evolution of cooperation in this game, see Maynard Smith (1982) and Lipman (1983). Another possibility is that the stakes could vary from one move to the next (Axelrod 1979). Still another is that more choices could be available to the players than the simple binary choice of cooperation or defection.

3. The interaction could involve more than two players at a time. The provision of collective goods provides the paradigm case of the n-person Prisoner's Dilemma (Olson 1965). Applications include a wide variety of problems in which each participant has an incentive to be a free rider on the efforts of others. Examples include the organizing of lobbying activities, and contributions to collective security. As Dawes (1980) has pointed out, the n-person case is qualitatively different from the two-person case in three ways. First, the harm caused by a defection is diffused over many players rather than focused on one. Second, behavior may be anonymous in n-person games. Third, each player does not have total reinforcement control over all the other players since the payoffs are determined by what many different players are doing. The literature on this is huge, but good places to start are Olson (1965), G. Hardin (1968), Schelling (1973), Taylor (1976), Dawes (1980), and R. Hardin (1982).

4. The abilities of a player to discriminate and to retaliate might each cost something. Therefore if nearly everyone else were using nice strategies it might pay to give up the abilities to discriminate and retaliate. This could help account

for the occasional atrophy of retaliatory capabilities. It could therefore provide a way of studying arms control and disarmament based upon evolutionary principles rather than formal treaties.

5. A player might not be certain about the choice actually made by the other player in the previous move. There could be problems of random noise or systematic misperception (Jervis 1976). To study this, the first round of the tournament was rerun with a 1 percent chance of misperception of the other's previous move. This resulted in yet another victory for TIT FOR TAT. This result indicates that TIT FOR TAT is relatively robust under conditions of moderate error in perception.

Chapter 9. *The Robustness of Reciprocity*

1. The Prisoner's Dilemma is slightly more general than this discussion suggests. The Prisoner's Dilemma formulation does not assume that the cost of helping is the same whether the other player cooperates or not. Therefore, it employs the additional assumption that both players prefer mutual help to an even chance of exploiting and being exploited.

2. Not surprisingly, successful executives in Washington learn to rely on reciprocity in this "government of strangers" (Heclo 1977, pp. 154–234).

3. The average scores of entrants include all of the rules except RANDOM, and take into account that the first round had 200 moves per game, while the second round had games of differing lengths which averaged 151 moves per game.

Appendix A. *Tournament Results*

1. The program for REVISED STATE TRANSITION contained an error and consequently did not always perform as intended. It did, however, serve well as a representative strategy by providing an interesting challenge for the other entries.

2. Here is how the hypothetical tournament scores are calculated. To make the constituency of a given representative five times as large as it actually was, let $T' = t + 4cs$ where T' is the new tournament score, T is the original tournament score, c is the coefficient in the regression equation of the representative whose effect is to be magnified, and s is the score of the given rule with that representative. It should be noted that the idea of a "constituency" of a representative is defined in this way, and that a typical rule is part of the constituency of several representatives. The hypothetical tournament in which the residuals are given added weight is constructed in an analogous manner with $T' = T + 4r$, where r is the residual in the regression equation for the score of a given rule.

Appendix B. *Proofs of the Theoretical Propositions*

1. To be precise, $V(B|B)$ must also be specified in advance. For example, if B is never the first to defect, $V(B|B) = R/(1-w)$.

BIBLIOGRAPHY

Acton, Thomas. 1974. *Gypsy Politics and Social Change: The Development of Ethnic Ideology and Pressure Politics among British Gypsies from Victorian Reformism to Romany Nationalism.* London: Routledge & Kegan Paul.

Alexander, Martin. 1971. *Microbial Ecology.* New York: Wiley.

Alexander, Richard D. 1974. "The Evolution of Social Behavior." *Annual Review of Ecology and Systemics* 5:325–83.

Allison, Graham T. 1971. *The Essence of Decision.* Boston: Little, Brown.

Art, Robert J. 1968. *The TFX Decision: McNamara and the Military.* Boston: Little, Brown.

Ashworth, Tony. 1980. *Trench Warfare, 1914–1918: The Live and Let Live System.* New York: Holmes & Meier.

Axelrod, Robert. 1970. *Conflict of Interest, A Theory of Divergent Goals with Applications to Politics.* Chicago: Markham.

———. 1979. "The Rational Timing of Surprise." *World Politics* 31:228–46.

———. 1980*a*. "Effective Choice in the Prisoner's Dilemma." *Journal of Conflict Resolution* 24:3–25.

———. 1980*b*. "More Effective Choice in the Prisoner's Dilemma." *Journal of Conflict Resolution* 24:379–403.

———. 1981. "The Emergence of Cooperation Among Egoists." *American Political Science Review* 75:306–18.

Axelrod, Robert, and William D. Hamilton. 1981. "The Evolution of Cooperation." *Science* 211:1390–96.

Baefsky, P., and S. E. Berger. 1974. "Self-Sacrifice, Cooperation and Aggression in Women of Varying Sex-Role Orientations." *Personality and Social Psychology Bulletin* 1:296–98.

Becker, Gary S. 1976. "Altruism, Egoism and Genetic Fitness: Economics and Sociobiology." *Journal of Economic Literature* 14:817–26.

Behr, Roy L. 1981. "Nice Guys Finish Last—Sometimes." *Journal of Conflict Resolution* 25:289–300.

Belton Cobb, G. 1916. *Stand to Arms.* London: Wells Gardner, Darton & Co.

Bethlehem, D. W. 1975. "The Effect of Westernization on Cooperative Behavior in Central Africa." *International Journal of Psychology* 10:219–24.

Betts, Richard K. 1982. *Surprise Attack: Lessons for Defense Planning.* Washington, D.C.: Brookings Institution.

Black-Michaud, Jacob. 1975. *Cohesive Force: Feud in the Mediterranean and Middle East.* Oxford: Basil Blackwell.

Blau, Peter M. 1968. "Interaction: Social Exchange." In *International Encyclopedia of the Social Sciences,* volume 7, pp. 452–57. New York: Macmillan and Free Press.

Bogue, Allan G., and Mark Paul Marlaire. 1975. "Of Mess and Men: The Boardinghouse and Congressional Voting, 1821–1842." *American Journal of Political Science* 19:207–30.

Boorman, Scott, and Paul R. Levitt. 1980. *The Genetics of Altruism.* New York: Academic Press.

Brams, Steven J. 1975. "Newcomb's Problem and the Prisoner's Dilemma." *Journal of Conflict Resolution* 19:596–612.

Buchner, P. 1965. *Endosymbiosis of Animals with Plant Microorganisms.* New York: Interscience.

Calfee, Robert. 1981. "Cognitive Psychology and Educational Practice." In D.C. Berliner, ed., *Review of Educational Research,* 3–73. Washington, D.C.: American Educational Research Association.

Caullery, M. 1952. *Parasitism and Symbiosis.* London: Sedgwick and Jackson.

Chase, Ivan D. 1980. "Cooperative and Noncooperative Behavior in Animals." *American Naturalist* 115:827–57.

Clarke, Edward H. 1980. *Demand Revelation and the Provision of Public Goods.* Cambridge, Mass.: Ballinger.

Cyert, Richard M., and James G. March. 1963. *A Behavioral Theory of the Firm.* Englewood Cliffs, N.J.: Prentice-Hall.

Dawes, Robyn M. 1980. "Social Dilemma." *Annual Review of Psychology* 31:169–93.

Dawkins, Richard. 1976. *The Selfish Gene.* Oxford: Oxford University Press.

Downing, Leslie L. 1975. "The Prisoner's Dilemma Game as a Problem-Solving Phenomenon: An Outcome Maximizing Interpretation." *Simulation and Games* 6:366–91.

Dugdale, G. 1932. *Langemarck and Cambrai.* Shrewsbury, U.K.: Wilding and Son.

Elster, Jon. 1979. *Ulysses and the Sirens, Studies in Rationality and Irrationality.* Cambridge: Cambridge University Press.

Emlen, Steven T. 1978. "The Evolution of Cooperative Breeding in Birds." In J. R. Kreps and Nicholas B. Davies, eds., *Behavioral Ecology: An Evolutionary Approach,* 245–81. Oxford: Blackwell.

Eshel, I. 1977. "Founder Effect and Evolution of Altruistic Traits—An Ecogenetical Approach." *Theoretical Population Biology* 11:410–24.

Evans, John W. 1971. *The Kennedy Round in American Trade Policy.* Cambridge, Mass.: Harvard University Press.

Fagen, Robert M. 1980. "When Doves Conspire: Evolution of Nondamaging Fighting Tactics in a Nonrandom-Encounter Animal Conflict Model." *American Naturalist* 115:858–69.

The Fifth Battalion the Cameronians. 1936. Glasgow: Jackson & Co.

Fischer, Eric A. 1980. "The Relationship between Mating System and Simultaneous Hermaphroditism in the Coral Reel Fish, *Hypoplectrum Nigricans* (Serranidae)." *Animal Behavior* 28:620–33.

Fisher, R. A. 1930. *The Genetical Theory of Natural Selection.* Oxford: Oxford University Press.

Fitzgerald, Bruce D. 1975. "Self-Interest or Altruism." *Journal of Conflict Resolution* 19:462–79 (with a reply by Norman Frohlich, pp. 480–83).

Friedman, James W. 1971. "A Non-Cooperative Equilibrium for Supergames." *Review of Economic Studies* 38:1–12.

Geschwind, Norman. 1979. "Specializations of the Human Brain." *Scientific American* 241 (no. 3):180–99.
Gillon, S., n.d. *The Story of the 29th Division.* London: Nelson & Sons.
Gilpin, Robert. 1981. *War and Change in World Politics.* Cambridge: Cambridge University Press.
Greenwell, G. H. 1972. *An Infant in Arms.* London: Allen Lane.
Gropper, Rena. 1975. *Gypsies in the City: Cultural Patterns and Survival.* Princeton, N.J.: Princeton University Press.
Haldane, J. B. S. 1955. "Population Genetics." *New Biology* 18:34–51.
Hamilton, William D. 1963. "The Evolution of Altruistic Behavior." *American Naturalist* 97:354–56.
———. 1964. "The Genetical Evolution of Social Behavior." *Journal of Theoretical Biology* 7:1-16 and 17–32.
———. 1966. "The Moulding of Senescence by Natural Selection." *Journal of Theoretical Biology* 12:12–45.
———. 1967. "Extraordinary Sex Ratios." *Science* 156:447–88.
———. 1971. "Selection of Selfish and Altruistic Behavior in Some Extreme Models." In J. F. Eisenberg and W. S. Dillon, eds., *Man and Beast: Comparative Social Behavior.* Washington, D.C.: Smithsonian Press.
———. 1972. "Altruism and Related Phenomena, Mainly in Social Insects." *Annual Review of Ecology and Systemics* 3:193–232.
———. 1975. "Innate Social Aptitudes of Man: An Approach from Evolutionary Genetics." In Robin Fox, ed., *Biosocial Anthroplogy,* 133–55. New York: Wiley.
———. 1978. "Evolution and Diversity under Bark." In L. A. Mound and N. Waloff, eds., *Diversity of Insect Faunas,* pp. 154–75. Oxford: Blackwell.
Harcourt, A. H. 1978. "Strategies of Emigration and Transfer by Primates, with Particular Reference to Gorillas." *Zeitschrift für Tierpsychologie* 48:401–20.
Hardin, Garrett. 1968. "The Tragedy of the Commons." *Science* 162:1243–48.
Hardin, Russell. 1982. *Collective Action.* Baltimore: Johns Hopkins University Press.
Harris, R. J. 1969. "Note on 'Optimal Policies for the Prisoner's Dilemma.'" *Psychological Review* 76:373–75.
Hay, Ian. 1916. *The First Hundred Thousand.* London: Wm. Blackwood.
Heclo, Hugh. 1977. *A Government of Strangers: Executive Politics in Washington.* Washington, D.C.: Brookings Institution.
Henle, Werner, Gertrude Henle, and Evelyne T. Lenette. 1979. "The Epstein-Barr Virus." *Scientific American* 241 (no. 1):48–59.
Hills, J. D. 1919. *The Fifth Leicestershire 1914–1918.* Loughborough, U.K.: Echo Press.
Hinckley, Barbara. 1972. "Coalitions in Congress: Size and Ideological Distance." *Midwest Journal of Political Science* 26:197–207.
Hirshleifer, Jack. 1977. "Shakespeare vs. Becker on Altruism: The Importance of Having the Last Word." *Journal of Economic Literature* 15:500–02 (with comment by Gordon Tullock, pp. 502–6 and reply by Gary S. Becker, pp. 506–7).
———. 1978. "Natural Economy versus Political Economy." *Journal of Social and Biological Structures* 1:319–37.
Hobbes, Thomas. 1651. *Leviathan.* New York: Collier Books edition, 1962.
Hofstadter, Douglas R. 1983. "Metamagical Themas: Computer Tournaments of the Prisoner's Dilemma Suggest How Cooperation Evolves." *Scientific American* 248 (no. 5):16–26.

Hosoya, Chihiro. 1968. "Miscalculations in Deterrent Policy: Japanese-U.S. Relations, 1938–1941." *Journal of Peace Research* 2:97–115.

Howard, Nigel. 1966. "The Mathematics of Meta-Games." *General Systems* 11 (no. 5):187–200.

———. 1971. *Paradoxes of Rationality: Theory of Metagames and Political Behavior.* Cambridge, Mass.: MIT Press.

Ike, Nobutaka, ed. 1967. *Japan's Decision for War, Records of the 1941 Policy Conferences.* Stanford, Calif.: Stanford University Press.

Janzen, Daniel H. 1966. "Coevolution of Mutualism between Ants and Acacias in Central America." *Evolution* 20:249–75.

———. 1979. "How to be a Fig." *Annual Review of Ecology and Systematics* 10:13–52.

Jennings, P. R. 1978. "The Second World Computer Chess Championshps." *Byte* 3 (January):108–18.

Jervis, Robert. 1976. *Perception and Misperception in International Politics.* Princeton, N.J.: Princeton University Press.

———. 1978. "Cooperation Under the Security Dilemma." *World Politics* 30: 167–214.

Jones, Charles O. 1977. "Will Reform Change Congress?" In Lawrence C. Dodd and Bruce I. Oppenheimer, eds. *Congress Reconsidered.* New York: Praeger.

Kelley, D. V. 1930. *39 Months.* London: Ernst Benn.

Kenrick, Donald, and Gratton Puxon. 1972. *The Destiny of Europe's Gypsies.* New York: Basic Books.

Koppen, E. 1931. *Higher Command.* London: Faber and Faber.

Kurz, Mordecai. 1977. "Altruistic Equilibrium." In Bela Belassa and Richard Nelson, eds., *Economic Progress, Private Values, and Public Policy,* 177–200. Amsterdam: North Holland.

Landes, William M., and Richard A. Posner. 1978*a.* "Altruism in Law and Economics." *American Economic Review* 68:417–21.

———. 1978b. "Salvors, Finders, Good Samaritans and Other Rescuers: An Economic Study of Law and Altruism." *Journal of Legal Studies* 7:83–128.

Laver, Michael. 1977. "Intergovernmental Policy on Multinational Corporations, A Simple Model of Tax Bargaining." *European Journal of Political Research* 5:363–80.

Leigh, Egbert G., Jr. 1977. "How Does Selection Reconcile Individual Advantage with the Good of the Group?" *Proceedings of the National Academy of Sciences, USA* 74:4542–46.

Ligon, J. David, and Sandra H. Ligon. "Communal Breeding in Green Woodhoopes as a Case for Reciprocity." *Nature* 276:496–98.

Lipman, Bart. 1983. "Cooperation Among Egoists in Prisoner's Dilemma and Chicken Games." Paper presented at the annual meeting of the American Political Science Association, September 1–4, Chicago.

Luce, R. Duncan, and Howard Raiffa. 1957. *Games and Decisions.* New York: Wiley.

Luciano, Ron, and David Fisher. 1982. *The Umpire Strikes Back.* Toronto: Bantam Books.

Lumsden, Malvern. 1973. "The Cyprus Conflict as a Prisoner's Dilemma." *Journal of Conflict Resolution* 17:7–32.

Macaulay, Stewart. 1963. "Non-Contractual Relations in Business: A Preliminary Study." *American Sociological Review* 28:55–67.

Manning, J. T. 1975. "Sexual Reproduction and Parent-Offspring Conflict in RNA Tumor Virus-Host Relationship—Implications for Vertebrate Oncogene Evolution." *Journal of Theoretical Biology* 55:397–413.

Margolis, Howard. 1982. *Selfishness, Altruism and Rationality.* Cambridge: Cambridge University Press.

Matthews, Donald R. 1960. *U.S. Senators and Their World.* Chapel Hill: University of North Carolina Press.

Mayer, Martin. 1974. *The Bankers.* New York: Ballantine Books.

Mayhew, David R. 1975. *Congress: The Electoral Connection.* New Haven, Conn.: Yale University Press.

Maynard Smith, John. 1974. "The Theory of Games and the Evolution of Animal Conflict." *Journal of Theoretical Biology* 47:209–21.

———. 1978. "The Evolution of Behavior." *Scientific American* 239:176–92.

———. 1982. *Evolution and the Theory of Games.* Cambridge: Cambridge University Press.

Maynard Smith, John, and G. A. Parker. 1976. "The Logic of Asymmetric Contests." *Animal Behavior* 24:159–75.

Maynard Smith, John, and G. R. Price. 1973. "The Logic of Animal Conflicts." *Nature* 246:15–18.

Mnookin, Robert H., and Lewis Kornhauser. 1979. "Bargaining in the Shadow of the Law." *Yale Law Review* 88:950–97.

Morgan, J. H. 1916. *Leaves from a Field Note Book.* London: Macmillan.

Nelson, Richard R., and Sidney G. Winter. 1982. *An Evolutionary Theory of Economic Change.* Cambridge, Mass.: Harvard University Press.

Nydegger, Rudy V. 1978. "The Effects of Information Processing Complexity and Interpersonal Cue Availability on Strategic Play in a Mixed-Motive Game." Unpublished.

———. 1974. "Information Processing Complexity and Gaming Behavior: The Prisoner's Dilemma." *Behavioral Science* 19:204–10.

Olson, Mancur, Jr. 1965. *The Logic of Collective Action.* Cambridge, Mass.: Harvard University Press.

Orlove, M. J. 1977. "Kin Selection and Cancer." *Journal of Theoretical Biology* 65:605–7.

Ornstein, Norman, Robert L. Peabody, and David W. Rhode. 1977. "The Changing Senate: From the 1950s to the 1970s." In Lawrence C. Dodd and Bruce I. Oppenheimer, eds., *Congress Reconsidered.* New York: Praeger.

Oskamp, Stuart. 1971. "Effects of Programmed Strategies on Cooperation in the Prisoner's Dilemma and Other Mixed-Motive Games." *Journal of Conflict Resolution* 15:225–29.

Overcast, H. Edwin, and Gordon Tullock. 1971. "A Differential Approach to the Repeated Prisoner's Dilemma." *Theory and Decision* 1:350–58.

Parker, G. A. 1978. "Selfish Genes, Evolutionary Genes, and the Adaptiveness of Behaviour." *Nature* 274:849–55.

Patterson, Samuel. 1978. "The Semi-Sovereign Congress." In Anthony King, ed., *The New American Political System.* Washington, D.C.: American Enterprise Institute.

Polsby, Nelson. 1968. "The Institutionalization of the U.S. House of Representatives." *American Political Science Review* 62:144–68.

Ptashne, Mark, Alexander D. Johnson, and Carl O. Pabo. 1982. "A Genetic Switch in a Bacteria Virus." *Scientific American* 247 (no. 5):128–40.

Quintana, Bertha B., and Lois Gray Floyd. 1972. *Que Gitano! Gypsies of Southern Spain.* New York: Holt, Rinehart & Winston.
Raiffa, Howard. 1968. *Decision Analysis.* Reading, Mass.: Addison-Wesley.
Rapoport, Anatol. 1960. *Fights, Games, and Debates.* Ann Arbor: University of Michigan Press.
———. 1967. "Escape from Paradox." *Scientific American* 217 (July):50–56.
Rapoport, Anatol, and Albert M. Chammah. 1965. *Prisoner's Dilemma.* Ann Arbor: University of Michigan Press.
Richardson, Lewis F. 1960. *Arms and Insecurity.* Chicago: Quadrangle.
Riker, William, and Steve J. Brams. 1973. "The Paradox of Vote Trading." *American Political Science Review* 67:1235–47.
Rousseau, Jean Jacques. 1762. *The Social Contract.* New York: E. P. Dutton edition, 1950.
Rutter, Owen, ed. 1934. *The History of the Seventh (Services) Battalion The Royal Sussex Regiment 1914-1919.* London: Times Publishing Co.
Rytina, Steve, and David L. Morgan. 1982. "The Arithmetic of Social Relations: The Interplay of Category and Network." *American Journal of Sociology* 88:88–113.
Samuelson, Paul A. 1973. *Economics.* New York: McGraw-Hill.
Savage, D. C. 1977. "Interactions between the Host and Its Microbes." In R. T. J. Clarke and T. Bauchop, eds., *Microbial Ecology of the Gut,* 277–310. New York: Academic Press.
Schelling, Thomas C. 1960. *The Strategy of Conflict.* Cambridge, Mass.: Harvard University Press.
———. 1973. "Hockey Helmets, Concealed Weapons, and Daylight Saving: A Study of Binary Choices with Externalities." *Journal of Conflict Resolution* 17:381–428.
———. 1978. "Micromotives and Macrobehavior." In Thomas Schelling, ed., *Micromotives and Macrobehavior,* 9–43. New York: Norton.
Scholz, John T. 1983. "Cooperation, Regulatory Compliance, and the Enforcement Dilemma." Paper presented at the annual meeting of the American Political Science Association, September 1–4, Chicago.
Schwartz, Shalom H. 1977. "Normative Influences on Altruism." In Leonard Berkowitz, ed., *Advances in Experimental Social Psychology,* 10:221–79.
Sheehan, Neil, and E. W. Kenworthy, eds. 1971. *Pentagon Papers.* New York: Times Books.
Shubik, Martin. 1959. *Strategy and Market Structure.* New York: Wiley.
———. 1970. "Game Theory, Behavior, and the Paradox of Prisoner's Dilemma: Three Solutions." *Journal of Conflict Resolution* 14:181–94.
Simon, Herbert A. 1955. "A Behavioral Model of Rational Choice." *Quarterly Journal of Economics* 69:99–118.
Smith, Margaret Bayard. 1906. *The First Forty Years of Washington Society.* New York: Scribner's.
Snyder, Glenn H. 1971. "'Prisoner's Dilemma' and 'Chicken' Models in International Politics." *International Studies Quarterly* 15:66–103.
Sorley, Charles. 1919. *The Letters of Charles Sorley.* Cambridge: Cambridge University Press.
Spence, Michael A. 1974. *Market Signalling.* Cambridge, Mass.: Harvard University Press.
Stacey, P. B. 1979. "Kinship, Promiscuity, and Communal Breeding in the Acorn Woodpecker." *Behavioral Ecology and Sociobiology* 6:53–66.

Stern, Curt. 1973. *Principles of Human Genetics.* San Francisco: Freeman.
Sulzbach, H. 1973. *With the German Guns.* London: Leo Cooper.
Sutherland, Anne. 1975. *Gypsies, The Hidden Americans.* New York: Free Press.
Sway, Marlene. 1980. "Simmel's Concept of the Stranger and the Gypsies." *Social Science Information* 18:41–50.
Sykes, Lynn R., and Jack F. Everden. 1982. "The Verification of a Comprehensive Nuclear Test Ban." *Scientific American* 247 (no. 4):47–55.
Taylor, Michael. 1976. *Anarchy and Cooperation.* New York: Wiley.
Thompson, Philip Richard. 1980. "'And Who Is My Neighbour?' An Answer from Evolutionary Genetics." *Social Science Information* 19:341–84.
Tideman, T. Nicholas, and Gordon Tullock. 1976. "A New and Superior Process for Making Social Choices." *Journal of Political Economy* 84:1145–59.
Titmuss, Richard M. 1971. *The Gift Relationship: From Human Blood to Social Policy.* New York: Random House.
Treisman, Michel. 1980. "Some Difficulties in Testing Explanations for the Occurrence of Bird Song Dialects." *Animal Behavior* 28:311–12.
Trivers, Robert L. 1971. "The Evolution of Reciprocal Altruism." *Quarterly Review of Biology* 46:35–57.
Tucker, Lewis R. 1978. "The Environmentally Concerned Citizen: Some Correlates." *Environment and Behavior* 10:389–418.
Valavanis, S. 1958. "The Resolution of Conflict When Utilities Interact." *Journal of Conflict Resolution* 2:156–69.
Wade, Michael J., and Felix Breden. 1980. "The Evolution of Cheating and Selfish Behavior." *Behavioral Ecology and Sociobiology* 7:167–72.
The War the Infantry Knew. 1938. London: P. S. King.
Warner, Rex, trans. 1960. *War Commentaries of Caesar.* New York: New American Library.
Wiebes, J. T. 1976. "A Short History of Fig Wasp Research." *Gardens Bulletin* (Singapore):207–32.
Williams, George C. 1966. *Adaptation and Natural Selection.* Princeton, N.J.: Princeton University Press.
Wilson, David Sloan. 1979. *Natural Selection of Populations and Communities.* Menlo Park, Calif.: Benjamin/Cummings.
Wilson, Edward O. 1971. *The Insect Societies.* Cambridge, Mass.: Harvard University Press.
———. 1975. *Sociobiology.* Cambridge, Mass.: Harvard University Press.
Wilson, Warner. 1971. "Reciprocation and Other Techniques for Inducing Cooperation in the Prisoner's Dilemma Game." *Journal of Conflict Resolution* 15:167–95.
Wintrobe, Ronald. 1981. "It Pays to Do Good, But Not to Do More Good Than It Pays: A Note on the Survival of Altruism." *Journal of Economic Behavior and Organization* 2:201–13.
Wrangham, Richard W. 1979. "On the Evolution of Ape Social Systems." *Social Science Information* 18:335–68.
Yonge, C. M. 1934. "Origin and Nature of the Association between Invertebrates and Unicellular Algae." *Nature* (July 7, 1939) 34:12–15.
Young, James Sterling. 1966. *The Washington Community, 1800–1828.* New York: Harcourt, Brace & World.
Zinnes, Dina A. 1976. *Contemporary Research in International Relations.* New York: Macmillan.

INDEX

Index

Index